# Revealing Statistical Principles

Jim Lindsey
*Department of Biostatistics*
*Limburgs University*
*Belgium*

A member of the Hodder Headline Group
LONDON • SYDNEY • AUCKLAND
Co-published in the USA by Oxford University Press Inc., New York

First published in Great Britain in 1999 by
Arnold, a member of the Hodder Headline Group,
338 Euston Road, London NW1 3BH
**http://www.arnoldpublishers.com**

Co-published in the United States of America by
Oxford University Press Inc.,
198 Madison Avenue, New York, NY10016



*British Library Cataloguing in Publication Data*
A catalogue record for this book is available from the British Library

*Library of Congress Cataloging-in-Publication Data*
A catalog record for this book is available from the Library of Congress

ISBN 0 340 74120 1

1 2 3 4 5 6 7 8 9 10

Printed and bound in Great Britain by JW Arrowsmith, Bristol

# Contents

# Preface

> To find out what happens to a system when you interfere with it you have to interfere with it (not just passively observe it). (Box, 1966)

This little book is addressed to people who are called upon to organize research studies involving human subjects or to judge the value of such studies, but who have little or no statistical knowledge. Those responsible for research and development in government and industry will be one main body of readers. Beginning Ph.D. candidates who will be conducting empirical research, requiring statistical methods, in some substantive area involving human beings but who have little training in such methods will form a second group.

In this book, I attempt to provide the basic *principles* of statistics in a non-mathematical way, accessible to a wide audience. My intention is to avoid technical details that can be obtained, as necessary, from a professional statistician or, for the more advanced, from the statistical literature. In this way, you should acquire sufficient knowledge of what statisticians do in order to be able to communicate with them, whether to obtain advice or to criticize the work they have done.

The text begins at the point when a study is originally conceived and moves in order through all stages to the final report writing, covering both observational and experimental (intervention) studies. Due to the primordial importance of the proper design of a study, much of the material concentrates on this aspect. I have spent considerably less time on the analysis, which, in any case, is covered, more or less adequately, in introductory statistics courses.

Even if you are primarily interested in only one of observational and intervention studies, you should preferably read both of these chapters. Many principles are common to the two and the contrasts can provide you with illuminating insights, highlighted by the quote from Box given above concerning the perpetual problem of studying causality among human beings.

The ideas presented in this book have accumulated from two types of experience: in educational planning and evaluation beginning about 25 years ago, primarily in Third World countries, especially India, Indonesia, Madagascar, and Morocco, and in clinical trials, beginning somewhat more recently and restricted to Europe. Thus, many of the examples in the text are related to these two fields, but I have tried to keep the discussion general enough to be applicable to any studies directly involving human beings and requiring statistical procedures.

In order to make the ideas clear and easily accessible, I have presented many as check-lists. My intention is not to provide many details on modern statistical methods but rather an overview. The bibliography will give you indications as to where to find some of the information needed actually to carry out the procedures that I have described.

Obviously, few of the ideas in this book are new. However, to make the text more readable, I have not loaded it with scholarly references, but have included in the bibliography the works that I have found most useful.

Philippe Lambert, Patrick Lindsey and six referees provided many useful comments on an earlier draft of various chapters.

I would like to acknowledge the support of UNESCO, which financed a course on this subject for the Ministry of Education in Morocco, and especially Claude Tibi who organized the course and who himself participated as much as I in presenting it.

J.K.L.
Diepenbeek and Liège
July, 1998

# 1

# Planning a study

## 1.1 Why statistics?

### *1.1.1 Human variability*

This book is about rigorous ways of collecting scientific information about human beings. In such circumstances, *random variation* in observations makes statistical procedures necessary. If all people reacted in exactly the same way in all circumstances, it would be possible to demonstrate any relationship of interest simply by observing one individual. If the common cold always lasted exactly seven days and administration of a new medication to one person reduced it to five, we would know that the drug worked. If every student received the same score on a test, administering it to one child would tell us how difficult it was. Because this is not so, we must conduct studies involving groups of people. And measures of variability will be as important as will averages.

However, although some specific group of people will be of particular interest in a study, usually you cannot observe all members of that group. You must select a representative subgroup, or *sample*. Thus, the field of statistics can provide you with objective means of generalizing from the particularities of observations only on some suitably chosen subgroup to conclusions about the group as a whole.

### 1.1.2 Research projects

We can divide most research projects concerning human beings into three main phases:

1. deciding on the question(s) to study, the procedures to use, and which people to include (Chapters 1, 2, and 3);
2. collecting the required information from and/or about them (Chapters 2 and 3);
3. processing, analysing, interpreting, and reporting this information (Chapters 4 and 5).

In a process of this complexity, detailed prior planning is essential. A considerable body of theoretical statistical knowledge is available to aid you in carrying out the first and third stages efficiently. Appropriate ways of performing the second are still very much a matter of trial and error, often depending on specific-subject matter questions rather than general statistical principles.

Statistical methods have a number of advantages over other methodological tools of the research worker. You must:

1. record information in as standardized a form as is possible with human subjects;
2. choose subjects in an objectively representative fashion so that you can make generalizations from specific observations;
3. state assumptions clearly, and usually check them empirically.

In planning a study, two aspects that will be of particular pertinence to statistics are that the results of the study are relevant to the questions being asked and that they are sufficiently precise. Statistical *design* of a study is specifically concerned with these objectives. General questions that you will have to face include:

- how to select the particular individuals to be observed;
- how to fix the total number of such individuals;
- how to allocate these individuals among various pertinent groups.

Thus, it is essential that a statistician be involved from the very initial stages of planning a study, and not simply be called upon to analyse the final results. In the latter case, unless one is particularly gifted or lucky, the statistician will generally only be able to provide a *post mortem* report on the reasons why the study failed to attain its goals!

It is important that you clearly distinguish between two types of investigation:

1. a planned *intervention* in the natural course of events to determine its effect;
2. the passive observation of phenomena as they exist in society.

The first is called an *experiment* or a *trial* and the second a *survey*. As we shall see, only the former can provide you with direct objective information about the consequences of the implementation of some innovation.

### 1.1.3 Ethics

Statistical methods can play an important role both in decision making and in scientific inference. However, they also have the potential for misuse: everyone is familiar with some phrase such as '*lies, damned lies, and statistics*'! It is best to consider some of these problems immediately.

Misuse of statistical methods can occur from the design, through the analysis, to the reporting stage of a study. Three of the most important problems to avoid are:

1. bias;
2. sampling too few subjects to detect a difference;
3. lack of published results.

The whole point of almost any study is to further knowledge, often with the view to using the information as a partial basis for policy or decision making. If you do not report the result of a study, it was a waste of time both for the investigators and for the subjects involved.

On the other hand, if it becomes evident to you during a study that there are unexpected difficulties, implying serious inadequacies, you should stop the study. Prior ignorance of the design and organizational requirements of a study is not an excuse for inadequate preparation!

One may argue (Altman, 1991, pp. 477–478, 491–492) that such misuse of statistics, and the accompanying substandard research, is unethical. You are:

- misusing subjects by exposing them to inconvenience and, in some cases, to risks;
- wasting resources;

- publishing misleading results that can lead to inappropriate decisions, with the accompanying further risks and wasted resources.

If you publish a poorly conducted study, others may:

- find it impossible to obtain funding or permission to conduct further research on the subject;
- be led to follow false lines of investigation;
- use the same inferior research methods elsewhere;
- widely introduce an intervention although it has no effect, or even harmful effects.

Unfortunately, with enough effort, even the worst research report can eventually be published somewhere.

Dishonesty and fraud are hopefully rare. Cases include:

- hoax – reporting a phenomenon that has never existed;
- forgery – inventing observations that were never made;
- cooking – selecting only those observations, or those statistical analyses, that agree with the desired conclusions.

Most cases of fraud are eventually uncovered, although sometimes only after significant damage has been caused.

## 1.2 Protocols

When beginning a study, you must develop a *protocol* to describe the purpose of the study and the steps in obtaining and analysing the data pertinent to this goal. However, there is no point in starting to plan a study that does not have adequate financial support and sufficient skilled staff available.

Proper design of the study is essential. The data from a well-planned study can be analysed in many appropriate ways, but no amount of clever manipulation at the analysis stage can compensate for a badly conceived study. A perfect design that is impossible to implement in practice is of no use; neither is a practically convenient plan that will not support the desired scientific conclusions.

Your protocol should clearly specify the following.

1. The subject:
   (a) the *background* and motivation;
   (b) the *question(s)* you wish to investigate;
   (c) the administrative *responsibilities*.
2. The material:
   (a) the *population* and *time frame* you will consider, and the *unit of observation* (person, family, town, ...);
   (b) how you will choose the *sample*, including the type of *study design*, *randomization*, the sampling or experimental unit(s), and the determination of *sample size*;
   (c) in experimental trials, the type of *subject consent*;
   (d) what *outcome(s)*, to become the response variables in the statistical analysis, you will measure;
   (e) what *sources of explanation*, to become the explanatory variables in the statistical analysis, you will measure.
3. The methods:
   (a) if you will maintain certain variables under experimental control, the randomization process by which

you will perform *treatment assignment* to individuals in the study;
    (b) what *instruments* you will use to measure the variables and how you will train the investigators involved to use them;
    (c) the ways that you will conduct *monitoring* of the progress of the study, including means of preventing deviations from the protocol and any interim analyses of the data.
4. The analysis:
    (a) procedures for *data transfer* to electronic form and for *verification* against recording and transcription errors;
    (b) appropriate *statistical models* that you think will allow you to detect patterns of interest in the data to be collected;
    (c) *selection strategies* for choosing among the possible models;
    (d) criteria to distinguish random, or chance, variability from that which is systematic.
5. The report: the form in which you will submit the final results.

These points form a unified whole; you should consider them simultaneously. Your choices about any point will have an impact on most of the others.

Prepare a draft protocol very early in your planning of a study. This will reveal confusions, weak points, and possible difficulties that you must face and resolve. You may require several drafts before producing a protocol that is acceptable on scientific, organizational, and ethical grounds.

Where possible, invite all people or organizations who may make use of the information obtained to provide input as to the structure of the protocol. In this way, they will be aware of the nature of the study and can make suggestions for modifications before the study begins.

It may often be desirable to construct the protocol in such a way that it will make the study comparable with previous existing studies, whether in the same or in other countries. This will be especially true for defining the population and constructing the instruments.

Involve the statistician who will be responsible for the analysis and presentation of the results from the first planning stages. It may also be necessary to consult with a statistical expert in study design.

To make the choices necessary to construct a protocol (except for the first point), prior knowledge of variability in the population and of ways in which it is practical to collect the information will be of help. For this, a *pilot study* may be necessary.

The final protocol will serve, among other things, as:

- a specification of the scientific design, including motivation and aims;
- an operations manual by which all investigators know what is expected of them;
- a prior record of assumptions and hypotheses so that you cannot be accused of drawing *post hoc* conclusions.

A protocol may vary from a few to 50 pages, depending on the complexity of the study.

### 1.2.1 Study validity

The final role of any study is to *convince*. Thus, in conducting a study, it is essential that people reading your final report are prepared to accept that your conclusions are *valid*. *Internal validity* refers to the extent to which your conclusions apply to the people actually studied, whereas *external validity* refers to the possibility of generalizing such conclusions to a wider population, whether persons, settings, or times. You must expect that the recipients of the final report will closely and critically question all of these.

#### Internal validity

**Relationship validity** The first and most fundamental type of questioning will be whether the relationships that you have found between the outcome and the sources of explanation are valid. Threats to this can come, for example, from claims that the relationships simply arose by chance in the sample examined or that the way in which you collected the data was biased.

Major problems may arise from:

- biases in study design or implementation;
- too much random variation in measurements;
- the sample size being too small to detect a relationship, called lack of *power*;
- applying an inappropriate statistical analysis;
- a relationship not being stated in the protocol, but found by 'data dredging' after the data have been collected.

**Causal validity** If the relationships found can be accepted, your second claim may be that they are causal. Causal relationships are most easily studied in closely controlled circumstances, but this will limit generalization of conclusions. Challenges to causal conclusions can arise, for example, if some other source of explanation can be invoked as influencing the outcome or if it can be argued that causality could be in the opposite direction. This can always occur if you have not used an intervention in the study and if you have not randomized the subjects involved to the treatments (Sections 1.4.1 and 3.1.3).

**Instrument validity** Even if the critics can accept the existence of a relationship, and, if applicable, the fact that it is causal, they may claim that the empirical phenomena observed do not correspond to the theoretical concepts proposed. In other words, the causal relationship that you have found is not what you claim it to be. Problems of instrument validity can arise from the ways in which you have measured either the outcome or the sources of explanation. It may be as simple as a bias, but may be related to complex problems of measuring attitudes and opinions (Sections 1.2.4 and 2.3).

### External validity

If the relationships you have found within a sample of people can be accepted, whether claimed to be causal or not, you must then ask to what extent they are generalizable to other people, in the same time and location, or elsewhere. If there is an interaction between a source of explanation and the type of subject, the setting of the study, or the time of the study, generalization of the results will be questionable. The essential technique here is the random selection of a sample from an appropriately and widely enough defined population of interest (Sections 1.4.1 and 2.1.2).

Experimental trials, because they apply interventions by using random allocation of treatments but involve no random selection of subjects, will generally have high internal validity but questionable external validity. In contrast, sample surveys, with their random sample selection but no treatment allocation, will have high external validity but no causal validity. Thus, when human beings are involved, internal and external validity are often in conflict. The use of strict control and homogeneity within a study will allow you to detect relationships more easily but will restrict the breadth of application of your conclusions.

### 1.2.2 Question investigated

The first step in preparing a protocol is to translate your vague general objectives, that have made a study necessary, into more detailed and specific objectives. This may entail developing working hypotheses that you can empirically test by the study. For all of this to become operational, you will usually have to choose some specific observable *outcome* as the principal object of study to be explained: being cured of a disease or becoming enrolled at school, for example. Such an outcome may be more or less 'natural', but you should always carefully construct and define it in an appropriate way. Be wary of predefined administrative or common-sense categories.

A number of steps are generally useful in developing the

central theme of the study:

- Search the literature to find other similar studies already available, whether in the same or other countries.
- Study the appropriate literature to discover the most suitable techniques of design and analysis for such a research project.
- Meet the people concerned to discuss the means of operationalizing all aspects of the study.
- Plan the budget carefully to ensure that the objectives can realistically be met.
- Where necessary and possible, consult outside experts.

### *1.2.3 Population and time frame*

Always try to have in mind a clearly defined population about which you plan to obtain information and a time frame to which it will be applicable. This is a complex technical question that I shall discuss in detail below and in the following chapters.

If, as is usually the case, you cannot study the whole population, you must also clearly specify the means of objectively choosing a representative sample of the appropriate size. This is at the centre of the design of the study. The principal designs will be described below.

In experimental trials, you can decide on the treatments or procedures to be compared. In observational studies, such flexibility is not possible. Having decided on the types of comparisons to make, search for some environment in which it is possible to collect data to provide such comparisons. Often, you must make do with comparisons that are far from ideal.

## *1.2.4 Instruments and measurements*

Besides the principal outcome to be explained, you will want to study the conditions under which it is produced (Section 1.3.3). This will require a careful operational definition of the *sources of exposure* that could explain the observed differences of outcome. If certain such conditions are to be under the control of the investigators, clearly define the means of assigning them to the subjects.

### Instruments

The protocol must specify the instruments that will be used to make the measurements, both of the outcome and of the sources of exposure, as well as the investigators who will use them. When necessary, it must also give the means of appropriate training of these investigators. Remember that a measurement, whatever the instrument, involves many known and unknown implicit theoretical variables, as well as unproven assumptions.

Three criteria are generally required for the evaluation of any instruments to be used.

1. *Validity:* Several types of instrument validity are important, although some are much more difficult to judge than others.
   (a) *Criterion validity* involves assessing an instrument against some accepted absolute standard.
   (b) *Construct validity* refers to whether the empirical phenomena being observed actually correspond to the theoretical concepts you wish to study. You can assess it by inspection of the pattern of relationships

between the instrument and other measures made in completely different ways.

(c) *Face* or *content validity* involves checking if the instrument (usually a questionnaire) covers the range of topics for which it is intended. A panel of experts usually makes the judgements.

These are all connected to the internal validity, so that they are also prerequisites for the external validity of the study as a whole.

2. *Reliability:* An instrument is *reliable* if it is able to yield the same results on repeated application. You may sometimes be able to accomplish this by looking at internal reliability at a single administration. Thus, for example, split-test reliability involves splitting your instrument (usually a questionnaire) into equal halves and checking the degree of agreement. The alternative is test-retest reliability, but take care that subjects do not change in any important ways between the two administrations. You should also assess reliability of results among different investigators using the instruments. Do this in normal operating conditions, because intensive training or special expertise will bias the results.
3. *Sensitivity:* The instrument should be able to detect scientifically important differences, or changes over time. On the other hand, you may waste money, and perhaps time, if you use overly precise instruments.

## Data recording

Common problems in data recording include:

- unclear specification of the data to be recorded;

- values need to be calculated instead of being entered as observed (for example, age from date of birth);
- too much data collected from each person;
- poor quality of recorded data;
- data recorded in a form unsuitable for transfer to a computer.

In collecting the data, it is better to anticipate problems than simply to wait for them to occur. Record *all* departures from protocol.

Together, this and the preceding subsection constitute what are classically called the *material and methods*.

### 1.2.5 Analysis and reporting

Arrange for all results to be directly produced in, or transfered to, a form that is machine readable. At this stage, verify all of the data to identify errors, cleaning and correcting as necessary.

Here, we are principally concerned with studies that require statistical analysis. You need only set out the main lines of such analyses in the protocol; these should be fairly flexible. Two main phases will be involved:

1. selecting among all possible statistical models those which are most appropriate to describe the patterns of interest in the data;
2. providing measures of precision of the unknown quantities in these models that are calculated, or *estimated*, empirically from the data.

Your choices will depend primarily on the type of outcome you have chosen for study.

Although the operations of statistical analysis are one of the cheapest aspects of a complete study, the time required to carry them through is often the most underestimated aspect of a study. Masses of unanalysed, and hence wasted, data that cost a great deal to collect lie stocked throughout the world.

Finally, you must report the results obtained in a form that is understandable by the audience to whom it is addressed. For these results to be convincing, your report must cover a clear description of all steps of the study to provide evidence that you carried it out in an objective and complete manner.

### 1.2.6 Monitoring the study

Follow your study closely to ensure that all aspects of the protocol are respected. Monitoring will be particularly important if:

- there is an intervention;
- there are several centres collecting data;
- the study extends over a considerable period of time.

Monitoring can serve a number of other functions as well, including:

- finding errors in reporting, if data are being entered in the computer as they are recorded;
- sustaining motivation by providing preliminary general results, called *interim analysis*;
- if an intervention is involved,
  - detecting adverse side effects;

– allowing the study to stop early if the intervention proves either ineffective or very effective.

Interim analysis is particularly delicate in any study. If you disclose partial results, this may influence future responses still to be recorded.

### *1.2.7 Administration*

Carefully plan the project management. This will include:

- Who has overall responsibility for the project?
- Who is in charge of various areas of the work, possibly divided both geographically and by subject?
- How are the various activities to be coordinated?
- What is the timetable?

You will have to establish a detailed budget covering:

- staff salaries;
- travel and subsistence;
- consumables, including general running costs and materials;
- equipment;
- overheads.

Before going ahead, be sure that you will have adequate funding available.

## 1.3 Types of observations

### *1.3.1 Choice of subject areas*

A basic problem is to select the most relevant items of information or types of observations from all those that it is practical to collect and that might conceivably have a bearing on the subject you are investigating. You may take a number of steps to resolve this problem:

1. Determine the details of the information required to deal with the problem.
2. Consider whether there are any related problems of importance on which this information, possibly supplemented to some extent, would throw light.
3. With the whole field mapped out in this way, consider the practicality of obtaining the necessary information covering any (sub)set of these problems.
4. Take final decisions on the inclusion of each point in light of the relative importance of the problems and the total load possible to impose on the investigators and on the subjects who will be involved.

The items of information that you will collect should form a rounded whole, covering a coherent area of interest.

You will only be able to collect accurate information if you obtain the full and willing cooperation of the investigators and of the subjects. Your study should have a clear purpose that you can explain to them, and the material that you collect must be relevant to this purpose.

You can distinguish three main types of observations that you will frequently need in a study:

1. objective facts;
2. opinions, attitudes, and motivations;
3. personal knowledge.

These will generally require different means of data collection and will be used for different purposes. If the observation unit is not human beings but, say, groups, only the first is usually pertinent.

### *1.3.2 Outcomes*

Your primary observation on each subject will be that of the phenomenon or outcome to be explained. The statistician calls this the *response variable*.

We may distinguish the *prevalence* of the phenomenon from its *incidence*. The former is concerned with the study of all existing cases that have the characteristic in which we are interested, as compared to those who do not. It refers to the *probability* of a case in the population whereas the latter is concerned only with new cases, referring to the *rate*, *risk*, or *intensity* of its occurrence.

Response variables can take a number of forms that will determine how the data are to be analysed:

- In many fields, the most common outcome type is *binary*, taking only two values, such as yes or no. Then, the phenomenon studied is the proportion of units in the population in each of the two categories: the proportion of children attending school or of people who are cured.
- A second common type of phenomenon to explain involves a *count* of something: the number of times a child has failed at school or the number of infections.

- Finally, a quantitative *measurement* may be made. Two forms of such measurements can be distinguished:
  - The observations can take any value, positive or negative. This is the main type of observation treated in many classical statistics books where the bell-shaped normal curve is emphasized, but is rare in practice.
  - Only positive values are possible, for example, length or duration in time, such as survival or length of unemployment.

You may also record other types of responses, such as membership in one of a number of categories. These may be unordered or ordered, called respectively *nominal* and *ordinal*.

Take special care in the selection and construction of the response variable, because the success of your study depends upon it.

### 1.3.3 Sources of explanation

The second type of observations on each unit will be the characteristics or sources of exposure that hopefully will explain at least some of the differences in the observed values of the response variable. The statistician calls these the *explanatory variables*. However, care must be taken with this term, because research workers in many disciplines call such observable quantities the parameters. As we shall see (Section 4.3.1), this latter term has a very different sense for the statistician, leading to problems of communication and to misunderstandings.

Explanatory variables take two main forms:

- two or more qualitative categories that separate the population into subgroups, such as sex, marital status, and so on;
- measured quantities, such as income.

Such variables can be useful in three main ways:

- as descriptive categories among which the response varies, such as sex or geographical region (in most contexts);
- as explanatory, but unmodifiable, characteristics, such as, for adults, amount of formal education or childhood illnesses;
- modifiable explanatory factors, such as accessibility to a public facility.

Of course, modifying an explanatory factor is only useful if it is a *causal* factor, as discussed below.

The tendency is often to accumulate a vast number of explanatory variables, but judicious choice of a relatively small number is usually preferable for a number of reasons:

- The cost of data collection and analysis will otherwise be unduly increased.
- The time required for each respondent to provide the information should be limited so that you obtain reliable data.
- Large databases increase the risk of recording and management errors.
- If you collect a large number of explanatory variables, the data analyst will be overwhelmed, and probably will be obliged to ignore many of them.

- The number of explanatory variables showing relationships to the response variable just by chance will also be increased.

Among other things, statistical analysis serves to determine which explanatory variables appear to have links to the response variable of interest, and in what way, given the inherent uncertainty arising from the variability when only a sample from the population of interest is observed.

### 1.3.4 Confounding

Many factors usually influence a response of interest, not all of which can be investigated at any one time. Any factor along with which the response varies is called a *confounding variable*. If it is unequally distributed in the groups being compared, it will give rise to differences in the response among the groups, distorting the comparison under study. Consider, for example, alcohol consumption, smoking, and lung cancer. Smoking and drinking tend to vary together. Hence, one might be led to conclude that lung cancer is caused by drinking.

In an experimental trial, you have three weapons to handle extraneous variables not under your direct control:

1. strict regulation of experimental conditions to reduce the effects of such variables;
2. direct measurement of such covariates to allow for them by matching, blocking, or stratification (Section 3.1.3) in the analysis;

3. randomization of treatment assignment to make the average effect of confounding variables the same in all treatment groups.

In observational studies, only the second strategy (Section 2.5.2) is generally possible. Your choice of environment is limited by the availability of the comparisons to be made. By definition, treatments are not assigned, randomly or otherwise, in an observational study.

Thus, in an observational study, confounding variables can be controlled by:

- stratification;
- matching similar individuals;
- measurement of *concomitant explanatory variables.*

These will be further discussed below.

### 1.3.5 Accuracy and precision

You must design any study in such a way that you can actually attain your desired objectives. Here, the *accuracy* of the results is their lack of bias, that is, you are actually measuring what you want to study. You must distinguish this from the *precision* of the results, the range of values within which what you are studying is almost sure to lie, usually assuming that the measurements are accurate.

Inaccuracies result from systematic biases in the methods of collecting data, particularly from:

- the selection of the individuals to observe – *coverage error;*
- missing responses – *non-response error;*

- the mode of data collection, as for differences in response by mail, by telephone, and in person – *design error*;
- the validity of the instruments used to make the observations – *instrument error*;
- effects on response due to the way the instrument is administered – *investigator error*;
- the accuracy of the information provided – *respondent error*.

You can never improve accuracy once the observations are made. In addition, to measure any study biases, you will require data external to the study itself.

The precision will depend primarily on:

1. the intrinsic variability of whatever is being observed, this generally being relatively large for human subjects;
2. the number of individuals upon whom observations are made and, to a lesser extent, the number of observations per individual;
3. the actual design of the study;
4. the precision of the instruments used;
5. to a minor extent, the type of analysis performed.

The first three points determine the sampling precision.

The *standard error* is a crude measure of the precision of an estimate obtained in a study. It is a function of the variability of the population, as measured by the *variance* or its square root, the *standard deviation*, and of the sample size, decreasing as that size increases.

If the variance is $\sigma^2$, then the standard error is $\sigma/\sqrt{n}$, where $n$ is the sample size. Approximately one-third of the observable random variability in an estimate will be greater than the standard error and one-twentieth greater than twice the standard error.

An estimate will be biased with respect to the population of interest if the study fails to include certain units because of coverage or non-response errors. In the same way, the standard error can only measure variability among samples due to not including all of the population in the observed sample; it does not take into account non-coverage and non-response.

Although only providing a rough estimate of precision for almost all types of response variables, the standard error will be useful for calculating the size of a study.

Lack of accuracy immediately places in question the value of any results and conclusions, whereas lack of precision generally only increases the uncertainty surrounding the exact values calculated.

### 1.3.6 Missing values

Non-respondents are almost always different than those who agree to respond, although the amount of difference may vary among questions to be answered. Thus, if non-response is not restricted to a small proportion of the sample, no general validity can be claimed for *any* conclusions drawn. Make every effort to reduce the number of missing values. At the same time, forcing people to participate or to reply to specific questions can bias the results because answers will not be reliable, and perhaps not even relevant. Institute a rigorous system of dealing with the non-response problem from the outset of the study.

Non-response may involve all answers for a given individual or only some of the answers. Many of the reasons may be similar in the two cases. Keep all respondents in the study who supply at least an answer to one question. They can furnish information about reasons for non-response.

The bias of non-response is approximately proportional to the rate of non-response ($R$) times the difference in parameter value ($\phi_A - \phi_M$) between the group answering and the missing group. Thus, increasing the response rate does not necessarily reduce bias if the missing group becomes much more extreme so that the difference between the two groups, $\phi_A - \phi_M$, increases more rapidly than $R$ decreases. Note, however, that the situation is usually even more complex than this because both the response rate and the parameter value will differ among types of non-response: for example, not contacted, incapable of replying, and refusal.

Substituting other individuals for the non-respondents is usually a mistake because the replacements will resemble the respondents, not the missing ones. It is not sufficient to plan for a sample of 1000 when 800 are required and 20% are expected to be missing. This is in no way equivalent to a complete random sample of 800.

In repeated surveys, such as panels, and longitudinal experiments, reduction of non-response may be especially important because it will tend to increase progressively. A continually larger number of missing respondents, called *dropouts*, can indicate that something is wrong, so that the study should either be reorganized or abandoned. In surveys, collection of information about friends and relatives of the participants at the beginning of the study can be helpful in tracing those who disappear. However, in other cases, study of the dropping-out process may be important in its own right, as when it involves drug side effects. Then, you should not discourage it but allow it to proceed 'naturally'.

## 1.4 Study designs

Optimization of study designs has (wrongly) primarily been concerned with obtaining maximum sampling precision for the least cost. It generally ignores questions of bias arising from questionnaires, investigator training, and so on, only taking into account those connected with missing those members of the population of interest not in the sample. In choosing a design, you must make choices as to using resources to maximize response rates, improve instruments, and so on, as well as increasing sampling precision.

### 1.4.1 Population and sample

Any group of individuals that you wish to study will be called the *eligible population*. You must clearly define it in such a way that you know what individuals belong to it or, at least, so that you know if any given individual belongs to it. It may often be desirable to define the population in such a way that it will make the study comparable with previous existing studies, whether in the same or other countries.

Often the eligible population is a subset of some larger *source population*. In practical situations, the latter will contain four subgroups:

1. the eligible;
2. the adequately assessed ineligible;

3. the assessed but unclassifiable because of incomplete information; and
4. the unassessed, due to lack of resources, unavailability, and so on.

Thus, for example, special difficulties will occur if the population contains 'floating' elements such as the homeless or nomads. In certain cases, for reasons of expense, you may have to exclude them. This may sometimes be justified by their differing fundamentally from the rest of the population. If they are important, a separately constructed study may be necessary.

Once you have defined, and enumerated, your population of interest, your problems do not end. Certain members may not be accessible, perhaps because you cannot locate them or because they are incapable of or unwilling to participate in the study.

To have practical value, the results of a study will generally need to be applicable to subjects other than those in the eligible population, for example to those who will enter that eligible population in future years. Thus, you will aim to apply the results to some *target population*. In contrast to the other two populations, this one is usually not fixed.

In most cases, the population will be so large that you cannot possibly observe all of the individuals in it, whether because of time constraints, expense, or other reasons. Then, a *sample* is any subgroup of the population that you choose to observe. Thus, you will have a five-level hierarchy from the source population to the sample:

1. source population;

2. target population;
3. eligible population;
4. accessible population;
5. sample.

Your selection criteria for inclusion in the sample will determine the external validity of the results of a study completely, in so far as they are actually fulfilled, and the internal validity to a large extent.

Once you have clearly defined the population, you will have to make certain fundamental choices as to the appropriate *design* of the study. Several basic principles are common to all designs. Two of the most important the following: are

1. Randomize wherever possible to maintain objectivity.
2. Calculate the minimum necessary sample size so as not to waste resources.

Let us look at these in turn.

## Randomization

As we shall see, randomization is used in selecting a sample from a population (Section 2.1.2) and, when some intervention is involved, in assigning subjects to groups receiving the different treatments (Section 3.1.3).

The term 'randomness' is an everyday common-sense notion that does not generally agree with what statisticians mean by the term. It is often associated with the idea of haphazardness. This latter term rarely if ever corresponds to the truly random in the sense defined below but often simply means that an event has no obvious explanation. In statistical randomness, the probabilities of the various possible events under consideration are (in principle) exactly known, whereas they are not in the everyday usage.

A very long completely random sequence of digits has the following characteristics:

- Each digit occurs equally frequently.
- Adjacent digits and sets of digits are independent of each other, so that you cannot predict the following digits from previous ones.
- Reasonably long sequences show regularity, such as about 100 ones in a series of 1000 random decimal digits.

Randomness is thus a property of the whole sequence, or more exactly of the process that generated it. You cannot judge a shorter subsequence drawn from it in isolation as to its randomness without knowing its source.

Traditionally, tables of *random numbers* were used. Now, you will usually generate such numbers by statistical software on a computer. As we shall see, these are used in various aspects of the design of a study.

## Sample size

Sample size will largely determine the precision of your results. Always calculate it before beginning a study. The technical details will be given in the following chapters; see, particularly, Sections 2.6 and 3.3.4.

- If it is impossible for you to finance a sufficiently large sample, so that the precision will be too low to draw useful

conclusions, then you should probably abandon the idea of making such a study.
- If your planned sample size is too large for the precision required, you will unnecessarily waste resources.

In the ideal case, you should specify the precision and then calculate the corresponding sample size, but this is often not possible and you must use the reverse process, of calculating the precision for a feasible sample size.

However, sample size is not the only important determinant of cost to take into account. Maximization of response rates and improvements in instruments and investigators are also both important and costly. It is usually difficult to weigh the relative benefits of each.

Unfortunately, sample size, and the resulting precision, are easily measured so that effort is often concentrated on it, at the expense of biases from non-response and inaccurate answers. Ignoring the latter in your calculations can lead you to greatly overestimate accuracy and precision.

### *1.4.2 Types of designs*

A number of different basic organizations of a study are possible.

#### **Prospective designs**

In a *prospective design*, you sample individuals from a population and then follow them over a certain period of time, recording new events. In principle, the idea is to start with groups having different values of some important explanatory variables, that is, different sources of exposure, and to follow them to see if different distributions of the response, the outcome, result.

We can distinguish three cases:

1. In an *experimental trial*, such as the *clinical trial* often used in medical studies, you will randomly allocate the subjects to one of a number of different treatments before the following observations.
2. In a *follow-up study*, you will follow distinct groups with different exposures, called *prognostic* or *risk factors*, to determine if they finally give a different response.
3. In a *panel design*, sometimes misleadingly called a cohort design, you simply observe all variables repeatedly as they occur over time.

In a *cohort design*, people of given ages are followed (strictly speaking, a cohort consists of all of the age group). This may be either prospective or historical.

Follow-up and panel studies are important for their ease of ensuring representativity, at least at the beginning, before drop-outs occur. Panel studies also cover the time dimension in a population. You may make observations more or less continuously, for example, using diary cards, or at intervals of time, such as once a year. However, such studies will only provide you with information about evolution as is, without any intervention, whether voluntary or external.

Experimental trials have the big advantage of allowing a direct causal interpretation because you have applied an intervention, but the major limitation of being impossible in most human situations. Even when possible, ethical considerations mean that you can usually only enroll unrepresentative volunteers.

Most trials look for a difference between two or more interventions or treatments. A particularly difficult type of trial to conduct is the *equivalence trial*, where you wish to determine if a new treatment is equivalent to the existing one. Does a new medication or teaching strategy provide as good results as that currently in use? If your study is too small, it will be incapable of detecting a difference so that you may draw the wrong conclusions from too little information.

## Cross-sectional designs

In a *cross-sectional design*, you simply record all variables on observed sampled individuals at one given fixed point in time. You can use them to study the state of a given population, for example, the prevalence of some condition. They are the easiest type of design to ensure a representative sample, by randomization, but have the major handicap of lacking a time dimension.

Do not confuse a series of cross-sectional studies using the same questionnaire with a panel study. In the former, different people are involved each time, whereas, in the latter, the same are used.

One special type of cross-sectional design is sometimes used. Aggregate measurements of some characteristics are compared across population groups, usually geographically defined, in an *ecological design*. Thus, you might want to relate the success rate in schools to the class size without taking into account individual student and teacher characteristics.

## Retrospective designs

In a *retrospective design*, you choose subjects according to their response values, their outcomes, and then obtain the values of the explanatory variables, the exposures. Thus, in contrast to all of the previous designs, here the explanatory variables are subject to variability, whereas the responses may be fixed and known. Many cross-sectional studies are, in fact, retrospective, because many questions apply to past history.

The major advantage is the speed with which you can obtain results, whereas the major problem is distortion of information as you try to go further back into the past. For example, you generally cannot use such a design to assess prevalence: a currently representative sample is not representative in previous points in time because of differential mortality, and so on.

In certain circumstances, this may be the only design possible. In epidemiology, it is often only after the victims have appeared that the origins of an epidemic can be studied!

A *case–control* design is a special type of design, usually retrospective, where a number of cases having a given characteristic are available. You then match these with similar control subjects who do not have the characteristic in an attempt to distinguish influential, hopefully causal, factors that occurred in the past. This design is often used when one of the response events is uncommon, as for a rare disease, because a prospective study would require an enormous sample to obtain even a few individuals with the event. In such a situation, this design is highly efficient in terms of the number of subjects and the time required. But such studies are more difficult to design properly, especially because of the choice of control group, than are standard prospective studies.

### *1.4.3 Causality*

When you observe the individuals in a sample survey at a given point in time, one kind of information that you should obtain concerns what you might observe if you chose another sample from the same population. However, to do this, you must assume that no change is taking place in the population between the two sets of observations. This is static information. Even if you observe the same sample over several time points, the information you obtain only refers to the evolution at those time points. You may extrapolate into the future, but this will only yield valid results if all of the conditions of change remain fixed, as previously observed.

In contrast, operationally, *causality*, in a statistical context, implies that changing one (explanatory) variable will produce changes in the *distribution* of another (response) variable. This differs fundamentally from the inferences you can make from surveys where you must assume that the population remains the same or continues to evolve in the same way. You cannot empirically study causality simply by taking static samples from a population, even by following them over time. Notice that, with this definition, an explanatory variable such as sex could not be a cause.

Causality, as so conceived, is a group or collective, not an individual, effect: two interventions cannot generally be compared on the same individuals (certainly not simultaneously), but only on two different groups (for an exception, see Section 3.3.1). You then study differences in the distributions of groups of responses.

The cause of some effect is often not unique. Both better manuals and superior teaching can improve student performance; several different drugs may cure the same illness.

Causality also implies a time sequence: an effect cannot occur before its cause. Theory should specify some time interval within which the effect will occur or last for appropriate measurement to be possible. However, in many cases, this ordering may not be obvious. For example, many diseases have a considerable latency period before the symptoms appear. It may not be possible to eliminate events occurring during that period as potential causes because the time of the true onset of the disease is not known. Even if the effect of some cause is theoretically instantaneous, testing the relationship will require a temporal precedence of intervention before effect.

In pure science, one searches for the causes of a given effect. In strictly applied work, one asks if a given cause (treatment) will produce the desired outcome. However, the latter is usually the required method for empirical study even in pure research.

Thus, instead of taking a sample of individuals from the population and observing the values of the variables that they have, as in a survey, suppose that you can select the individuals and then control them by giving them values of the variable(s) that you think are causes. This type of planned intervention is called an *experimental trial*. As we have seen above, the assignment of such values is usually done randomly, for the same reasons as in choosing a sample from a population, especially to eliminate biases.

To perform an experiment, you must have:

- at least an approximate theory predicting the size, or direction, of effect of an intervention;
- a suitable group of subjects prepared to give consent to the intervention;
- means of (unrealistically) isolating the phenomenon studied from external sources of influence;
- stable responses whose only reason for changing over time is the treatment variable;
- measuring instruments whose resolution (precision) is fine compared to the size of the predicted effect.

In studies involving human beings, causality is thus very difficult to ascertain empirically, which is not to belittle its extreme importance. Think of the relationship between smoking and lung cancer. The debate lasted for many years, although sampling from existing populations showed a strong association whereby proportionally more smokers had lung cancer. But an experiment could not be performed in which some people were randomly chosen and told to smoke and others not, after which cancer incidence would be observed in the following years.

In such cases, where an intervention is not possible, the best plan is to attempt to discover as many different consequences as possible of the causal hypothesis under study. Thus, for example, with smoking and lung cancer, we could look at the death rate for:

- people smoking different amounts in the same time;

- those smoking the same amount but in different times;
- ex-smokers and current smokers of the same amount;
- ex-smokers smoking different amounts;
- ex-smokers smoking the same amount but stopping at different times in the past;

(see Cochran, 1965).

Take particular care when drawing conclusions from an ecological design. Such studies may provide clues to relationships among individuals but can suffer from the ecological fallacy. Suppose that you make a study in a set of groups or clusters, say geographical regions, and that you have available global measures of some response and a corresponding source of exposure for each cluster. Although the two measures vary together, this provides no direct evidence of any links between the response and the exposure at the individual level, one of the main reasons being that unavailable confounding factors could explain the relationship. Individual success at school may not be linked to class size even although the overall school success rate is; larger classes may be found in more deprived neighbourhoods, with individual success depending on social class.

Both in a static survey and in an experiment, you may find a relationship of dependence between two variables. The statistical procedure to describe the relationship may be the same in both cases. But your conclusions about the meaning of the relationship must depend on the way in which you collected the information. No mathematical manipulation of the data afterwards can change this. You can only directly study causality empirically if you can manipulate the appropriate explanatory variables. You can only draw causal conclusions from a survey, without experimentation, by making empirically unverifiable assumptions.

### *1.4.4 Choosing a design*

Many of the points already discussed in this chapter can be summarized by considering issues surrounding the choice of a study design.

The first question that you must decide in selecting a design is whether an intervention will be involved or not. Experimental trials have the enormous advantage of allowing causality to be empirically studied without untestable hypotheses. However, they often may require a very long study duration between intervention and effect. When a trial is ethically and logistically possible, it is usually preferable. However, most studies on human beings are not experimental.

It cannot be emphasized enough that not all kinds of designs will allow you to draw the same types of conclusions, in particular those about the causal effect of one variable on another. Only an experimental trial can answer such questions clearly.

Observational studies have the significant advantage of generally providing no added risk to the people involved. Among such studies, a *descriptive* survey is designed to estimate some simple characteristics of a population, whereas an *analytical* survey is to investigate associations among such characteristics.

Retrospective and prospective studies are both *longitudinal designs*. They can provide information about processes over time. However, in a prospective study, current practice may change in unforeseen ways over time, making the findings irrelevant.

With a prospective study, the sample can usually be clearly defined and chosen to be representative of the population of interest. This is often much more difficult for a retrospective study. However, the prospective study can be more subject to missing data, especially drop-outs. Retrospective studies are generally completed much more quickly and cost less, but are subject to increasing inaccuracy as you go back in time (unless suitable written records are available).

An ecological design is particularly important when the characteristics of interest are relatively homogeneous in each area and measurement errors on individuals are relatively large. Then, contrasts among regions, for example, among cultures, may provide the evidence you require. Thus, for example, in descriptive epidemiology, ecological evidence, by comparison among countries, has indicated links between diet and cancer.

## 1.5 Summary

Random variability in observations makes statistical procedures necessary. Statistics can help you in all stages from setting up a study to analysing and reporting it. If you expect to call upon a statistician for help in the analysis, involve her or him from the beginning of the design stage.

One of the most important distinctions is between making passive observations of subjects and an intervention. Only the latter will allow you to draw causal conclusions without making empirically unverifiable assumptions, but, in most situations, it is impossible to perform with human subjects for ethical or other reasons.

Do not begin any study without preparing a detailed protocol outlining all steps of the procedures to be followed. The two main types of observations that you will make are the outcomes to be explained and the corresponding sources of explanation. You must take into consideration many factors that will control their accuracy and precision. You can most easily manipulate sample size, but others, such as instrument biases and missing data, will generally be much more important.

In choosing the design of the study, definition of the population of interest is a first important step, followed by randomization where possible for all relevant aspects, especially in choice of sample and assigning intervention treatments. Designs may be retrospective, cross-sectional, or prospective, each with their particular advantages and disadvantages.

You must seriously consider all of these aspects of a study before beginning the actual data collection. When you have made the appropriate choices, you should state them clearly in your protocol, a document that will help you to justify the objectivity of your work when you make the final report.

# 2

# Sample surveys

## 2.1 Sampling

In observational studies, one of the first choices to make is whether to study the whole population or only a sample from it.

### *2.1.1 Samples versus censuses*

In extreme cases, where you require information on all individual units, you must make a complete *census* of the population. Generally, the cost in effort and expense required to collect information is less, per unit, for a census than for a sample. However, if the size of the sample needed to give the required precision represents only a small fraction of the total population, the total effort and expense required to collect information by sampling methods will be much less than that for a census.

A sample generally has a number of advantages over a census:

- A full census may be impractical because of the cost, time, and effort involved.

- You can much more easily ensure the completeness and accuracy of the results if you only collect information from a small proportion of the population. Generally, forms are more completely and more accurately filled in. Furthermore, you can make more detailed checks of the quality.
- You can obtain more detailed information about each unit in a sample, even with a smaller total volume of data.
- You can generally obtain results much more quickly by means of sampling than by a complete census. This is especially true at the stages of collection and recording.
- Sampling using interviewers is necessary in a population where many people are illiterate and could not fill out a census form.

The amount of information that you will obtain from a sample depends on its absolute size, not on its size as a proportion of the population, at least when the proportion is small.

Your final choice between a sample survey and a census will usually depend on which gives you the highest degree of precision and accuracy for the least cost. The latter will depend on a number of factors, including:

- the amount of information required per individual;
- the number of individuals to be covered;
- the way in which individuals are distributed in the region studied and their accessibility;
- the size of the region studied and the quality of the transportation and communication network;
- the type of instruments used;
- the qualifications and training of the investigators using the instruments.

You must carefully weigh all of these factors.

### 2.1.2 Random sampling

One can imagine choosing a sample from a population in a wide variety of ways:

- a readily accessible group;
- haphazard choice, most often used in experimental situations, where it is vaguely and implicitly assumed that items selected are typical;
- expert choice or judgement sampling of some representative members;
- volunteers, where specific changes of behaviour must be accepted, again most often in experimental contexts;
- quota sampling, used in opinion polls and market surveys, whereby the interviewers themselves build up a sample roughly proportional to the population on a few demographic variables.

All of these procedures have at least two major disadvantages:

1. they are always biased in unknown ways with respect to the population;
2. they do not allow any statistical calculation of precision of the estimates.

You will not be able reliably to generalize the results of a study based on any of these choices of sample to any known population.

Easily accessible groups are often unique. Experts rarely agree. Volunteers, by definition, are exceptional. Interviewers tend to select subjects who are easy to find, who are likely to be cooperative, or who they think may benefit from the study. In all cases, the bias is constant with sample size, never decreasing as more individuals are observed.

Usually, you will wish the sample to be 'representative' of the population; you want the individuals to be exchangeable, as far as possible, for all of their specific characteristics that are not of interest. You can only accomplish this by choosing a *random sample*: every member of the population, independently, has a known, non-zero probability of being selected for the sample. Thus, usually, you will require that the observations selected from the population be *independent*: observing one tells you nothing about which others may be selected. In contrast, with haphazard selection you do not know the probabilities of selection.

With random selection, you have the best chance of a reasonable and unbiased balance of the unknown characteristics, although this cannot be guaranteed.

### 2.1.3 Observational and sampling units

The *observational unit* is the entity about which you are directly collecting information. This does not mean that you must obtain all information from that unit. If you are studying children, you may require relevant information about their family, the school, the village, and so on. Often, the only way that you can obtain it accurately is directly from each such group.

The unit of observation may not be the same as the *sam-*

*pling unit*, which is the entity chosen at random from the population. Thus, the sampling unit might be the family, whereas the observational unit might be the eldest child or even all children in the family. Sampling units may not be of the same size, but may contain differing numbers of observational units. Several different levels of sampling units may be necessary in the same study, such as school, classroom, and child.

Generally, for a given *sample size* of observational units, the smaller the sampling unit employed, the more precise and representative the results will be. This is because observational units within a sampling unit tend to be similar, providing less information than independently chosen units. This requirement often conflicts with costs because larger sampling units are generally easier to observe.

If you are interested in the inter-relationships among members of a group, then collect information on such groups as a whole, or at least on pairs of units within such groups. Similarly, if you are interested in inter-relations among the behaviours of the same individuals at different time points, design the study so as to provide information over an adequate time period.

Thus, your choice of a sampling procedure will depend not only on the relative precision of the competing techniques but also on practical considerations. The most suitable method will depend on the type of information already available about the population. For example, you should not oblige the investigators to travel excessively and you should subject them to proper supervision and control.

### 2.1.4 Sampling frame

A *sampling frame* is a list of all sampling units in the population. This is necessary in order for you to be able to make a random selection. If no such frame already exists, its construction may constitute a sizeable part of the work of the survey.

Notice that it is not necessarily required that a list of all observational units be available: the population of sampling units and the population of observational units may not be identical. When they are not, a random sample of sampling units does not yield a random sample of the observational units.

A sampling frame may be defective in a number of ways:

- The information about the units may be inaccurate. Some units may not even exist at all.
- The information may be incomplete if certain units in the population are missing. These may be random individuals or whole categories of the population. The latter is much more serious.
- There may be duplication whereby certain units are included more than once.
- The sampling frame may be out of date, in that it was accurate, complete, and without duplications at the time of construction, but the population has changed.

You will generally be able to discover inaccuracies, and to correct them, as the study progresses, as, in many cases, you will find the duplications. On the other hand, you will not usually find the *coverage errors*, due to incompleteness;

these will often lead to some categories of the population being under-represented. Thus, the former is measurable whereas the latter is not.

## Sources of frames

You can obtain information to construct a list of sampling units in various ways.

- *Population census*: A complete census will tend to be out of date, especially because it can only be carried out infrequently and the results take a considerable time to be released.
- *Administrative lists*: Various administrative activities require lists of segments of the population. They will generally only be accurate, complete, and up to date if the administration is very efficient. Often, they are maintained by local offices so that their accuracy may vary throughout the country.
- *Lists of establishments*: schools or hospitals (for cluster sampling).
- *Lists of households or dwellings*: Such lists, for example for taxes or elections, have more permanence than lists of individuals.
- *Town plans and maps*: Considerable detail is required for these to be useful. Otherwise, you may need to include all dwellings in a block or other small region. Take care because equal-sized areas will not be equally densely populated. In rural regions, you may need to use natural boundaries, with unequal areas, because you cannot easily locate rectangles marked on a map on the ground.

- *Lists of villages*: If available maps are insufficient for rural areas, you may use villages as the sampling unit. Take special care if all individuals are not affiliated with villages.

Note that lists of individuals are not suitable if the sampling unit is a larger entity such as the household.

## 2.2 Organization

A number of the ways in which a particular society is organized can play important roles in the success of a survey:

1. common language(s) so that communication is possible;
2. common assumptions and understandings;
3. freedom for interviewers to contact sampled people;
4. lack of fear of strangers;
5. trust that answers will be held confidential;
6. belief that surveys are useful and informative.

Each can be a factor in non-response.

### 2.2.1 Types of surveys

We may distinguish three different types of studies:

1. collection of relatively simple facts covering the whole population of a country and capable of giving separate results for small administrative regions, usually done by a census;
2. surveys of the whole population of the country involving more detailed information, but not providing details for small regions;

3. local surveys covering a small region to obtain detailed information by field investigators.

The first type of study presents relatively simple sampling problems but may be administratively complex. The third type is usually simple both in sampling and administration. The second type is the most difficult, usually benefiting from a relatively complex sampling design.

### 2.2.2 Administration

Timing of the study will be influenced by:

- seasonal factors;
- availability of an appropriate sampling frame;
- holidays;
- deadlines for the results, and so on.

However, the principal requirement will be that you conduct it at a time representative of that to which you will apply the results. Once you establish the timing, you must set up a schedule for the various stages.

Costs must be estimated. These may include:

- expert consultation fees;
- sample selection;
- printing questionnaires;
- travel and subsistence expenses;
- data entry, verification, and analysis, including computers and software;
- preparing the report;
- general overheads, including salaries.

Reasonable estimates are necessary to ensure that the study is feasible.

The amount of administrative work will depend on:

- the scale of the study;
- the sample design;
- the area covered.

In the field, the main task will be to supervise the investigators, whereas centrally it will be to direct the data recording and analysis. Different people will often be responsible for each. If the area covered is large, you may require regional centres for the first task. Use existing administrative and office organizations where possible.

### 2.2.3 Ethical questions

A basic principle for any study should be that those involved should be part of a population that is in a position to benefit from the results of the study.

You must use ethical means to obtain any list of people to be used as the sampling frame. Many countries have data protection acts that limit access to lists of names and addresses. You may need to recontact for permission any people who have supplied information without being explicitly told that it would be used for research. This can incur substantial non-response even before data collection begins.

Any collection of information from individuals involves problems of confidentiality. Such private information is protected to differing degrees in different countries. Generally, you should safeguard the confidentiality of respondents at all costs. Longitudinal studies are particularly challenging; you will be collecting such a volume of information on each person that it may be easy to identify specific individuals.

Take special care that individuals cannot be identified if you are to release data banks for research purposes outside the organization collecting the data. You may have to remove certain information, such as geographical details. You may want to release 'restricted use' files to researchers who swear to abide by specific procedures to safeguard the security of the data under penalty of law.

A further ethical issue involves the amount of information about the goals of the study that you supply to the respondents. Hiding the true purposes of a study from the respondents may seem necessary in order to obtain honest answers in some circumstances, but raises difficult ethical questions that you must carefully weigh.

In many cases, you may have to call upon an ethics committee to make a decision about the various procedures used in your study.

### 2.2.4 Pilot studies

When you know little about the population or are using new and untested instruments, a small preliminary study will generally be necessary. A *pretest* is a piecemeal check of parts of the instrument, whereas a *pilot study* is a small test version of the full study.

A pilot study may aim to:

- check the adequacy of the sampling frame;
- develop the field procedure by
  - testing the adequacy of the questionnaires;

  - training the investigators;
  - checking the efficiency of the briefing and instructions for the investigators;
  - verifying communication between the field and the office;
- obtain information on the various components of variability to which the population is subject;
- determine the suitable size of sampling unit;
- estimate the rate of non-response, whether refusals or non-contacts;
- provide estimates of the costs, such as interview and travel times.

Where possible, try to use random sampling for the pilot study, although this is rarely done in practice. Instead, some typical sampling units, such as nearby villages, are generally selected. Take care that the members of any pilot study are not included in the final sample to avoid any biases arising from repeated interviewing.

Where possible, conduct the testing in two stages:

1. a trial of the questionnaire by professional investigators who are thoroughly familiar with it and with the study as a whole;
2. a subsequent trial using the revised questionnaire with investigators of the type who will actually be used in the study.

You may want to try open questions in the pretest, or pilot study, in order to determine the range of possible answers. From this, you can then construct closed questions (which does not mean that the final instrument should only contain closed questions). In certain cases, you may find it necessary to try several forms of the same question to check if different answers are received.

## 2.3 Measuring instruments

If you lack research or administrative experience in the subjects to be covered, it is fatally easy to omit some vital items when designing the instruments.

Where possible, it is generally preferable to use instruments that have already previously been used in other studies. This has several advantages:

- substantial time can be gained;
- the instruments have already been tested in a similar context;
- results of the study will be more comparable with other studies.

As already mentioned, thoroughly try out all instruments, in the conditions of the new study, by means of a small pretest of the survey procedures.

### *2.3.1 Types of instruments*

The purpose of an interview is to find out what is on someone's mind, to discover things that cannot be observed directly. You may use a large number of different instruments to collect information about the problem of interest. These include:

- direct observation schedules;

- tests of personal knowledge;
- questionnaires consisting of closed and/or open questions;
- structured interviews;
- recording undirected discourse (life histories);
- participant observation.

Notice that certain of these instruments, especially later in the list, often do not directly produce information appropriate for subsequent statistical analysis.

The more detailed the information you obtain from each observational unit, the smaller, and often the more unrepresentative, the sample will usually need to be. The more structured the information you obtain, the more suitable it will be for statistical analysis, as opposed to more subjective means of summarization.

You can apply many of these instruments in a variety of ways.

- Direct administration:
  - observation;
  - direct interviews;
  - telephone interviews.
- Indirect methods:
  - deposit the questionnaire to be collected later;
  - diary cards collected periodically;
  - postal surveys.

Among other things, your choice will depend on;

- the budget available;
- the number of observational units;

- the ability of people to reply with and without guidance;
- the interest generated by the study.

Postal surveys can be relatively cheap and can have high return rates. However, they do have a number of disadvantages over interviewing:

- The questions must be simple and clear, not requiring a lot of explanation.
- All answers are final, with no opportunity to attempt to overcome hesitation or ambiguity.
- Spontaneous replies, opinions uninfluenced by discussion with others, and tests of personal knowledge are impossible.
- Questions cannot be ordered such that early ones are answered without knowledge of later ones.
- There is no guarantee that the person randomly sampled actually answers the questionnaire.
- Supplementary observational data cannot be obtained.

Pretesting of the questionnaire is especially important with postal surveys because the investigator will not be present to gain cooperation or to clear up ambiguities. The covering letter and the sponsorship can be crucial in convincing people to reply.

Generally, where possible, direct observation is preferable to questions, and questions on facts and on past actions are preferable to questions on generalities and on hypothetical future actions. Physical measurements are more objective, but qualitative observations are often more capable of summarizing the important features of a complex situation. By proper standardization and calibration among investigators, you can make qualitative observations reasonably objective.

### Multicultural studies

Special problems arise if the instrument is to be used in several cultures. You can distinguish two cases:

1. The instrument may be adapted for use in more than one language and cultural context, without any attempt to compare the results cross-culturally.
2. The results of the study must be compared or aggregated across cultures.

In the first case, you should only use the original instrument as a guide in producing a culturally appropriate procedure for the new setting. The second case is much more complex, raising conceptual, technical, and ethical problems. In all cultures:

- both the items in the instrument and the responses to them must be conceptually and functionally equivalent;
- the same phenomenon must be measured – the underlying concept must exist and be pertinent;
- the questions must be relevant and not too personal or offensive;
- issues of local importance must not be missed.

In most existing instruments, the majority of items are highly culture-specific.

You should have your instrument translated into the target language, preferably in several alternative versions, and then back-translated to the original language for checking. Have a panel of monolingual lay people comment on the translation in the new language; multilingual people are never representative, especially if one of their languages is English. Psychological and emotional states are the most difficult to translate between cultures. When working in several cultures, you must use much more elaborate field pretesting for reliability and validity in each language.

### 2.3.2 Questionnaires

You may design a questionnaire for completion in three principal ways:

1. by the investigator,
   (a) from direct observation or
   (b) with the aid of questions put to the respondents;
2. by the respondent with little or no assistance from the investigator.

You may require all of them in the same survey for different types of information. In all cases, clearly specify the means of distinguishing a non-response from a non-applicable question.

The simplest instruments are those where the investigators themselves record observations. Generally, there should be a separate instruction booklet, so that the form itself remains simple.

When the investigators are to fill out the questionnaire containing questions posed to respondents, you may train them to use a given fixed wording or ask them to elicit information in any way that will provide an answer to the questions. Thus, you must instruct them as to whether they must put the questions in the exact form given or can ask them in a more general way. The former procedure is especially important in matters of opinion where the wording may affect the answer.

In such a context, explanatory notes may either be on the questionnaire or in a separate booklet. The latter results in a more compact questionnaire and is more suitable for professional investigators. The former is more likely to ensure that the investigators remain aware of the purpose of each question.

When the respondents will fill out the questionnaire themselves, the only role of the investigators may be to explain the purposes of the study and to persuade the respondents to cooperate. Be especially attentive to the wording of the questions and the explanatory notes. The latter will generally be on the questionnaire beside each question. However, detailed and lengthy explanations should be avoided.

The principal types of questions are:

- measurements, in clearly defined units;
- multiple-choice, or closed, questions, where all possible answers are fixed in advance, *one and only one of which must necessarily be chosen*;
- semi-open questions, with the main possibilities listed, but alternative replies can also be supplied;
- open questions, where any reply is possible.

Never record measurements in predefined categories, except in the rare cases of delicate subjects such as income, where an accurate answer is otherwise unlikely.

The advantage of multiple-choice questions is that they are usually easy to understand and to reply to and that they are also easy to record and to analyse. However, they carry the danger of telegraphing the answer and provide no opportunity for nuance. The most common error is to allow respondents to choose several possibilities; this makes statistical analysis difficult or impossible.

Semi-open questions may suggest to the respondent that the list provided contains more common or more acceptable answers than the one not included that might be closer to the truth.

Open questions, if well formulated, can provide information on almost any subject. They are the only possibility when you cannot predict most or all answers to a question. However, they can be more difficult to construct to obtain objective answers and much more difficult to record and to analyse.

Thus, pay careful attention to the detailed wording of all questions, even if these are only intended as a guide to the investigator, and to the order in which they are presented. In formulating questions, a number of basic principles are important: they should each

- be precise, simple and logical, with no technical terms unfamiliar to the population being addressed;
- be short, concerning only one idea;
- be unambiguous – for example, age can refer to the last or to the nearest birthday;
- be self-explanatory, where possible;

- not ask for too many details or about events too far in the past;
- require some answer, so that it is possible to detect non-response;
- clearly specify any units of measurement and the precision required;
- provide enough different categories, without any overlapping intervals;
- be answerable – asking for the cause of death of parents must be conditional on their being dead and not require excessive medical details;
- avoid hypothetical situations.

In summary, a good question will:

- be relevant to the respondent;
- be easily understood and unambiguous in meaning;
- not be influenced in any untoward way by the context in which it is used;
- relate to the survey objectives;
- mean the same thing to the respondent, the investigator, and the decision makers for whom you are performing the study.

For every question, ask yourself if it is really necessary. A shorter questionnaire reduces non-response and increases accuracy of answers.

The order of the questions is important. The logical order is not necessarily the best. The first questions should help to place the respondent at ease. Simple demographic questions often serve this purpose well. Then, the more delicate subjects can be treated near the middle. Generally group questions by subject areas so that the respondent is not required to change train of thought too often. An exception will be duplicate or redundant questions that you use to double-check each other and that must appear at separate moments in the sequence.

In complex cases, you may need several questionnaires for each sampling unit: child, household, village, school. Take great care in identification so that you can link them all correctly together for analysis.

The questionnaire forms must be convenient to use and set out so that the information can easily be translated to electronic media. Only collect raw data; the investigator should make no intermediate calculations. For example, dates are more reliable than time intervals. You should usually allow space for the investigator or respondent to make general comments. Although you cannot easily treat such observations by an exact analysis, they can be of considerable value in drawing your attention to relevant facts not covered by the questionnaire.

In all cases, arrange the information to be recorded on the questionnaire in such a way as to be easily transferred to a computer. Plan and clearly specify the manner in which to do this, including the systematic handling of non-responses, at the same time as you develop the questionnaire. No intermediate calculations should be necessary at this transfer stage.

### 2.3.3 Field investigators

The main tasks of a field investigator are:

- locating the people sampled;
- obtaining agreement to be interviewed;
- asking the questions;
- recording the answers.

In many cases, only about one-third of an investigator's time is actually spent interviewing. The rest involves:

- studying materials;
- travelling and locating respondents;
- editing questionnaires;
- general administrative work.

You must allow for this in scheduling work-loads.

Field work is very arduous, involving considerable mental strain. Interviews are often intense experiences involving complete attention and frequent thinking on one's feet. Hours of work are generally very irregular, because evening visits are often necessary to contact working people and reduce non-response. Supervisors should be required to undertake some field work themselves in order to be able to appreciate the difficulties.

Payment by piece rates is generally unsatisfactory, because it leads to incomplete or hasty work and to irregularities such as replacing one respondent by another.

If you are recruiting new investigators, give all applicants preliminary tests and arrange proper training courses. Recruits should be honest, interested, accurate, adaptable, pleasant, and able to follow complicated instructions. Carefully watch and supervise the early work of new investigators. If possible, build up the team of investigators by selecting suitable ones at the pilot stage; this provides a means of testing and training.

Investigators should have background knowledge of the subject under study. Unspecialized teams of investigators are only suitable for carrying out routine studies requiring relatively simple questionnaires. When a high degree of technical knowledge is needed, use staff in existing organizations.

Note also that the reactions of the respondents may depend on the origin of the investigators. For example, government officials may arouse suspicions that the information collected could be used for purposes other than those stated.

### 2.3.4 Accuracy of recorded information

Once you have chosen a proper sample, the most common sources of inaccuracies in the observations are variation in the respondents' reactions to the method of assessment and variation in the investigators' techniques. These measurement errors may be called *observation errors*. Inadequate responses may be missing, incomplete, irrelevant, or inaccurate. In order to be able to check if these problems are evolving over time, record the sequence in which the respondents are interviewed.

In addition to the principles listed above for questionnaire construction, you should consider a number of further points in judging if you will obtain accurate information from the respondents:

- Are the respondents sufficiently informed, or able to recall past events, to be capable of providing accurate answers?

- Can they translate into an unambiguous and understandable answer what they believe to be the truth, and are they willing to do this?
- If the answers require substantial work, will they be prepared to do it?
- Do they have motives for concealing the truth (perhaps simply trying to impress or please) and, if so, will they refuse to answer or give incorrect information?

A number of remedial actions are possible:

- Anonymity of respondents is important, especially when using incriminating or highly personal questions.
- If you are using skilled investigators, working on a relatively small sample, they may be able to elicit accurate information in many circumstances.
- In certain cases where you expect such problems to arise, you can take substitute measures, such as approximating revenue by the size of the dwelling.

In other cases, you may have to abandon the question.

Similar points apply to the role of the investigators:

- Are they sufficiently informed about the subject and motivated to do the work required?
- Are they in possession of standardized methods for eliciting and recording qualitative information in an objective manner?
- Are they trained to approach all respondents in the same way, independently of the opinions they hold and the disposition they happen to be in?

- Do they know how to avoid indicating to the respondent in any way what they believe to be the appropriate answer, instead listening with the required patience?
- Are they meticulous in recording the answers supplied or in verifying that the respondent has properly filled out a questionnaire?

Fieldwork will be most accurate if the investigators are well trained, capable, conscientious, and keen. Personal characteristics of the investigators can influence the answers given. Make field checks where possible. Carry these out on a random subsample of units, in a way such that the investigators know beforehand that they will be checked but not which parts of their work will be examined.

Poorly constructed instruments, badly administrated, can also lead to biases and inaccuracies, some already mentioned.

- Misinterpretation may occur because the question is not specific and the respondent does not want to show ignorance.
- Technical terms and academic jargon can easily lead to incomprehension.
- For multiple choices, an insufficient number of alternatives may force the respondent to choose an inappropriate answer.
- Providing the possibility of 'don't know' may allow an easy way out for lazy respondents but is necessary for people who genuinely do not have experience of a subject.
- Probing by the interviewer is usually acceptable for factual questions but not for opinions or tests of knowledge.

- Leading questions will cause some subjects to provide what they believe to be the correct or desired answer.
- People tend to choose an item near the beginning or end of a long list of possibilities.
- In a series of questions involving ratings, the first will often give more extreme results because the respondent has not yet established a standard.
- More recent information is usually more accurately reported. However, this may not always be the case. When asking for the number of events in the previous week, the answers may be, consciously or unconsciously, telescoped so that more are reported than actually happened. In such cases, reports of events in a longer period, such as a year, may be more accurate.

For binary variables, *sensitivity* is the proportion of those who actually have the characteristic who are classified as having it. *Specificity* is the proportion of those not having it correctly classified. Obviously, it is usually difficult to check either of these because you do not generally know the correct answer.

Make a preliminary examination of the returned forms as quickly as possible so that you can have defective work corrected while it is still possible.

## 2.4 Sampling error

If instruments could be constructed such that they could yield accurately recorded data, that is, no observation errors, observed differences in response among groups in a sample could arise from four sources:

1. the (causal?) effect of belonging to the group;
2. other relevant factors not taken into account, called confounding (Section 1.3.4);
3. bias in choosing the sample;
4. random chance in choosing the sample.

Your study has internal validity if you can ascribe differences to the first source. Now, let us consider the last two.

Two types of *sampling error* can arise:

1. *biases* in selection of the sampled units;
2. chance differences between members of the population included in the sample and those not included, called *random sampling error*.

Bias forms a constant component of the error that does not decrease, in a large population, as the number in the sample increases. On the other hand, random sampling variation, what statisticians have called random error, decreases on average as the sample size increases.

Bias is an important factor in determining the *accuracy* of the results, along with observation error. On the other hand, random sampling error determines the *precision* of any quantities estimated (Section 1.3.5). There is an inverse relationship between the latter two: precision increases as random sampling error diminishes.

### 2.4.1 Causes of sampling bias

Random selection is not haphazard selection. You can only obtain a true random sample by adhering to some strict random process (Section 1.4.1). Sticking pins on a map or meeting people on a street corner is not random. If at all possible, perform the random selection centrally, and do not leave it to the arbitrariness of the individual field investigators.

Faulty selection of the sample can give rise to bias in a number of ways. The main causes are:

- deliberate selection of a 'representative' sample (Section 2.1.2);
- a selection procedure depending on some characteristic that is associated with the properties of the units that are of interest – many haphazard selection processes have this defect, as when, in a shopping survey, customers arriving at shops are interviewed;
- conscious or unconscious deviations from a proper random selection process, for example when a field investigator replaces a sampling unit for some reason, such as substitution of a convenient member of the population when difficulties are encountered in obtaining information from the individual randomly selected;
- failure to cover the whole chosen sample, leading to missing data.

Of course, bias will also arise from any systematic measurement errors as discussed above, for example if the respondents misunderstand a question.

If possibilities of bias exist, you will not be able to draw any fully objective conclusions from a sample. The only universally acceptable way for you to avoid bias in the selection process is to draw the sample at random and to avoid missingness wherever possible.

## 2.4.2 Missing values

Missing values can arise from coverage errors or from non-response errors. The former are systematic for any sample chosen, whereas the latter depend on many factors related to the actual conduct of a given study. Only non-response can generally be detected. Let us consider it in more detail.

There are many reasons for non-response.

- A respondent may be unsuitable for interview, because of an error in the sampling frame.
- There may be difficulty in contacting respondents, which can depend on various factors:
  - change of residence;
  - the nature of the respondent, where, for example, housewives are more often at home than those who go out to work;
  - the time of call, with employed people being away during the day or in vacation time;
  - the interview situation, for example, if there is advance notice of the visit.
- Refusals may depend on:
  - the disposition of the respondent, this varying from cheerful cooperation to hostility;
  - the techniques of the investigator;
  - the number, nature, and sequence of the questions;
  - sponsorship of the survey.
- Respondents may lack of interest or concern – people with children in school will be more prepared to answer questionnaires about education and ill people more inclined to answer about health.

- They may have some incapacity or inability, such as
  - illness in the family;
  - language difficulties.

If non-response is not minimal, review both the questionnaire and the type of investigator to find out whether it is wise to continue. If you have not detected these problems during the pilot stage, the required changes will often render the earlier and later results incompatible.

You can fairly easily remedy two problems:

1. failure to make contact with the selected respondents – persistent calling back may be the only solution, although contacts with neighbours may provide useful information to trace the missing individuals;
2. too long and complex a questionnaire – shorten and simplify it (at the pilot stage!).

Making an appointment for an interview can have the perverse effect of allowing the respondent to be out at the time of call.

In a well-designed questionnaire, you will include a section to record information on non-respondents. For personal refusals, you should have a considerable amount of approximate information noted. Once the survey is finished, you should check the missing answers to see if they are representative at least for known groups, such as sex, age, and so on.

Missing answers to single questions can occur for a number of reasons:

1. the interviewer forgets to ask the question;

2. the respondent cannot provide an answer;
3. the respondent refuses to answer;
4. the answer is not recorded.

The rate of missing answers is often linked to interviewer experience.

Measures to reduce the non-contact and overall refusal rates, such as calling back, are often more costly than those to reduce individual missing items. The latter usually involve improving interviewer training and questionnaire design. You must weigh the cost of reducing non-response bias against increased sampling error due to the resulting smaller sample size for fixed total cost.

### 2.4.3 Random sampling error

Sampling error arises from non-observation: the whole eligible population, the sampling frame, is not included in the sample. Thus, the simplest way for you to reduce random sampling error, and increase precision, is to increase the sample size. Other things being equal, this error is approximately inversely proportional to the square root of the sample size. However, the precision attained also depends on the variability in the population. Techniques discussed below, that restrict selection as compared to full randomization without introducing bias, can increase precision. The main one is stratification.

If you only require overall results for the whole population, you can attain a given degree of precision with a far smaller sample than will be the case if you require the detailed results for different parts of the populations (for example, different regions, towns, and so on).

## 2.5 Sample designs

For any random sampling procedure to be possible, you must subdivide the population under study into sampling units. As we have seen, these may be the observational units, or some aggregation of them. They may be natural (families), administrative (villages), or artificial (square regions of equal area or population density).

### 2.5.1 Simple random samples

A simple random sample from the population of observational units is the simplest type of rigorous method of obtaining a sample. It is also the basis of most other procedures. In this method, you divide the population into observational units, a numbered list of which is available, and select the required number of units at random from this entire population.

Usually, you will generate random numbers by statistical software on a computer. The numbers you obtain indicate which units on the list you are to include in the sample. In simple random sampling, each member has the same probability of being selected. Although the simplest method, this is not usually the most efficient or the most cost-effective.

### 2.5.2 Stratification

If you have available additional information about all of the individuals in the population, you can obtain increased representativity and precision by *stratification* of the sample. This involves randomly choosing fixed proportions in each category of some known explanatory variable such as regions, age groups, or sex. In this way, the sample and population proportions are guaranteed to be equal, or to have a known relationship, at least for this variable.

Divide the population into blocks, or strata, of units, such that the members of each stratum are as similar as possible on some important criteria. These strata may or may not all contain the same number of units. You then sample each stratum at random.

The main purpose of stratification is to increase precision of the overall population estimates and of the corresponding estimates for each stratum. If there are large differences in response among the units in the various strata, the accuracy and precision of the overall estimates will be increased. This is because the strata will be represented in their correct (or at least known) proportions, whereas, in simple random sampling, these proportions are subject to random sampling error. In a stratified sample, only variation within strata produces random sampling error.

For stratification to be possible, you must be able to classify each sampling unit distinctly into one stratum. In other words, you must have complete information about any stratum variable available for all sampling units before beginning the study. Typically, the formation of only a few strata will yield the most gains.

In summary, stratification is only possible if the information about each sampling unit necessary to create the strata is available in the sampling frame. Stratification yields three main advantages over simple random sampling:

1. If units within a stratum are more similar with respect to

the response than those between strata, then the precision of any overall population estimate will be greater than that from a simple random sample of the same size.
2. The corresponding estimates within strata should be more accurate, which will be important if these subgroups are of special interest.
3. Stratification will make it possible to sample various subgroups in different ways, which may reduce costs.

Be careful about simplistic assumptions concerning the relationship between sample size and costs. This relationship may not be linear. For example, cost per unit may be a decreasing function of sample size in some strata.

### Uniform and variable sampling fractions

If you select the same proportion of the members in the population from each stratum, the strata will all be represented in the correct proportions in the complete sample. However, different proportions may be more useful if you allow the more important or more variable strata to be overrepresented. In this case, you may need to use appropriate weightings in subsequent calculations.

Note, however, that a known non-zero probability of *selection* does not imply a similar probability of *measurement*. Different groups of people may have different rates of nonresponse introducing differential biases.

Often, the largest strata are most variable, but also have the lowest sampling costs. If there is a fixed cost in each stratum, as well as a variable cost proportional to sample size, you will minimize total cost by choosing strata samples proportional to the product of strata size and strata variability (measured by strata standard deviations) divided by the square root of cost per unit.

In the opposite case, you may oversample small strata if it is important to have precise information about them.

### *2.5.3 Clustered or multi-stage samples*

In simple random and stratified samples, the sampling unit and the observational unit are identical. These designs are only possible when you have a complete sampling frame of the observational units available. In many countries of the world, they are not feasible for this reason, or because of high travel costs. Then, you can only apply random sampling to groups or *clusters* of observational units; these become the sampling units.

Thus, you may sometimes save time and expense, at the cost of reduced precision, by clustering, that is, by choosing random groups of individuals found together. Then, you may study all of members of each chosen group or select some randomly. For example, you may choose several people from each of several villages or entire classrooms of students, where you take the village or classroom at random. When the clusters are geographical regions so that the sampling frame is a map, this is called area sampling.

Individuals in such a cluster will generally be more similar or homogeneous than if you had chosen each of them independently randomly from throughout the whole population. This means that you are actually collecting less information for a given number of observations, resulting in lower precision per observation unit. However, in certain cases, you may increase overall precision, because the smaller cost per unit sampled allows you to take a larger overall sample for the same total cost.

With cluster sampling, the variance, as compared to a simple random sample of the same size, will be increased by a factor of approximately $1 + (m - 1)\rho$, where $m$ is the cluster size and $\rho$ is the correlation among responses in a cluster. Thus, in contrast to stratification, where strata should be as homogeneous as possible to increase precision, here clusters should be small and heterogeneous (have small correlation).

In *multi-stage sampling*, you consider the population to be made up of a number of first-stage sampling units. You then take those chosen to consist of second-stage sampling units, and so on. (Simple clustering is two-stage.) At each stage, you sample the units by a suitable method, usually simple random sampling or stratified sampling; this may not be the same at all stages. The important point is that you choose all sampling units at all stages by a proper random process.

Choice of the first-stage units is especially important. You should take into account several criteria:

- The total number of primary units in the population should be relatively large.
- The units should have clear boundaries. Well-known administrative units are often preferable.
- The units should be fairly uniform in size.
- The units should remain stable over time (at least from when the information was obtained until the survey is carried out). Comparability to past and future data is even better.

Multi-stage sampling has several important advantages:

- It introduces flexibility that is absent from the simpler methods. You can use existing natural divisions of the population as sampling units.
- You will only need to carry out subdivision into second-stage units on those first-stage units actually selected.

However, multi-stage sampling generally yields less precise results than a sample containing the same number of final-stage observational units selected by some suitable one-stage process.

Construction of an appropriate pilot study is usually most difficult for multi-stage sampling. You will require a much more extensive pilot study than for other designs if you want to obtain any reliable preliminary estimate of variability.

If you are interested in the clusters themselves, as well as their members, many of the optimizing criteria discussed in this section are no longer relevant.

### 2.5.4 Systematic samples

Much practical sampling is not fully random in nature. A frequent method of selecting a sample, when a list of all units is available, is to take every $k$th entry on the list. The first entry should be determined by selecting a random number between 1 and $k$.

Such a systematic sample would be a simple random sample if the list were arranged completely at random. You can only estimate precision if you make this assumption. However, no list is completely random. If you are forced to use this method, take great care to verify that no periodic features or monotonic trend appear in the list, especially any that might be associated with the sampling interval, $k$.

This method does have several advantages:

- It often involves less labour and technical expertise than a true random sample. Thus, it may be especially useful if selection must be made in the field by relatively untrained investigators.
- In some situations, you need not know the complete sampling frame in advance. You can select units sequentially in time.

### 2.5.5 Case–control designs

When the response characteristic that you wish to study is extremely rare in a population, you would require an impossibly large sample to obtain even a few cases randomly. If cases are available, but in a non-random way, you may need to use a radically different type of study. You identify subjects with the condition of interest, as well as a group of controls without the condition, matched as closely as possible to the cases. You then compare them as to their previous exposure to any risk factors of interest.

The main advantages of such a case–control study are practicality, simplicity, speed, and low cost. However, the disadvantages are many:

- non-random selection of the cases;
- difficulty in locating appropriate controls, as well as their non-random selection;
- inaccuracy of retrospective memory;

- differential recall between cases and controls;
- detection bias, whereby one of the exposure factors studied facilitates observing the response condition.

You must study cases carefully for conclusions to be generalizable. Estimation of exposure from a control group that includes individuals either predisposed to such exposure or not can be altered by changes in this mix. Identification of eligible cases and controls must not depend on their exposure status.

You may locate cases as all those diagnosed

- in a community;
- in a random sample from the population;
- in all relevant institutions (schools, hospitals) in the community;
- in one or more such institutions.

The controls should be as similar as possible to the cases, except that they do not have the condition being investigated. Without the possibility of randomization, this can be difficult or impossible. Often, the best chance is by individually *matching* one or more controls with each case on some variables that could confound the comparison. These should be variables that are strongly related both to the condition and to the exposure factors. But then you cannot use such variables as possible risk factors for the condition.

Problems of memory with retrospective observations have already been discussed. However, here there is a further problem. Subjects with the condition may have thought about the reasons for it and have noticed exposure factors. These may also be present for the controls, who, however, have not noticed them. Thus, exposure factors may be under-reported in the controls. Subjects exposed to certain factors may more often seek professional advice that, in turn, allows their condition to be more often detected than the unexposed, creating additional bias.

You must interpret any results from a case–control study with extreme caution.

### 2.5.6 Repeated sampling

Most samples are carried out on a single occasion to determine the characteristics of the population at a given point in time. If the population is subject to change, such a study cannot provide you with information on the nature or rate of change. One possibility is to work retrospectively, and ask questions about the past. This generally involves problems of memory, whereby information further back in time is less trustworthy.

In other cases, you must make provision to redo the study periodically, in what are called *waves*. This may take different forms. You may:

- develop a completely new survey for each point in time (these independent random samples may overlap);
- repeat the survey in the same form but with new sampling units drawn at intervals in time (again, the random samples may overlap);
- repeat the survey with the same sample at each time point, a panel or cohort study;
- replace a part of the sample at each occasion;

- draw a subsample from the original sample, and only resurvey these units.

Your choice will generally depend on the exact type of change to be studied, as well as on questions of cost and practicability.

Panel and cohort studies will provide you with the most information about how change is occurring. You may design one so that you choose fixed proportions of respondents in categories of some key explanatory variable, the *risk groups*. For example, to study lung cancer, you could select groups of smokers and non-smokers.

These studies will not require your respondents to remember events over long periods, thus increasing accuracy. However, they carry several dangers:

1. Respondents must be *tracked* from interview to interview, with the major risk of drop-outs; the people lost will not be representative.
2. A sample that you randomly selected to be representative of some population when the study began will usually no longer be representative at successive waves, if only because everyone has grown older and the younger cohorts are missing.
3. Repeated restudy of the same units may induce resistance to providing information, but it may also lead to more accurate answers, both creating spurious trends in the results.
4. It may result in modification of the units involved, as compared to the rest of the population, so that they become less representative.

5. If initial risk groups were sampled, the high-risk group may be followed more closely, leading to surveillance bias, or subjects may change habits, and hence risk groups, for example, by stopping smoking.

Recruiting people willing to be included in a study requiring continued participation over a period of time will often not be easy. One of the most difficult and costly aspects of longitudinal studies is keeping track of the respondents. To facilitate this, it is usually a good idea to collect as much information as possible about the respondents' families, close relatives, and friends. If you lose contact with respondents, you can contact these people to try to locate them.

The advantages of cluster designs, such as using hospitals or schools as sampling units, may be lost in longitudinal studies because the observational units may change clusters between waves. If you administer questionnaires in groups, this will also create an increasing problem as respondents become split up over time.

The sample for a longitudinal study is representative at the moment it is chosen. However, it may not remain so as time goes by. For example, a sample of students chosen in the sixth year of school will, two years later, no longer be representative of students in the eighth year of school because of failures and perhaps drop-outs. One possible approach is to use *freshening* by sampling extra students at each new time point in an attempt to bring the sample back to representativity. In such cases, it may also be useful to obtain retrospective background information on these supplementary respondents.

In general, diary cards are better than repeated interviews for registering new events or changes in circumstances. On the other hand, interviews allow in-depth questioning about chronic conditions.

## 2.6 Sample size

Sample size refers to the number of observational, not sampling, units. A sample size can only be calculated for some specific aspect of the population to be estimated, usually related to some important response variable. The size of sample that you will require in order to attain a given precision for such an estimate depends on the variability of the population and on the extent to which it is possible to reduce the different components of this variability in the random sampling error, primarily by stratification.

The standard error, although usually a crude measure of the precision of an estimate obtained from a sample, is accurate enough to allow you to make sample size calculations, these being, in any case, themselves rather rough. We have seen that the standard error is a function of the sample size, decreasing as that size increases (Section 1.3.5).

Suppose that, in simple random sampling, you want to be relatively confident that the population value of interest is within a small region around an estimate of it calculated from the sample. Then, you can use the standard error to calculate the approximate size of sample required to ensure this. Fortunately, in the common cases, the standard error is easy to calculate.

Let us represent the population value of interest by $\phi$, and its estimate calculated from the sample by $\hat{\phi}$. If the standard error is represented by $\sigma/\sqrt{n}$, where $n$ is the sample size, you wish to have sufficient precision from the sample to be confident that $\phi$ lies in a small interval around $\hat{\phi}$. Applying the criteria mentioned in Section 1.3.5, you can use the approximate interval, $\hat{\phi} \pm 2\sigma/\sqrt{n}$.

### 2.6.1 Binary responses

For a binary response variable, in the simplest case, you will be interested in the proportion of the population having a given characteristic, that is, in the *probability* of that characteristic appearing. Let us call this $\pi$, so that $\phi = \pi$ is the value of interest. Then, you can obtain the estimated standard error from

$$\hat{\sigma} = \sqrt{\hat{\pi}(1 - \hat{\pi})}$$

Thus, for the desired precision interval of two standard errors, say $\hat{\phi} \pm \delta/2$, to be small enough, you will require that

$$n = \frac{16\hat{\pi}(1 - \hat{\pi})}{\delta^2}$$

In order to calculate the sample size using this formula, you need to have some idea of what value the population proportion has — this is a major reason why sample size calculation is always approximate.

Suppose that you think that the proportion is about 80% and you want an interval of $\delta = 4\%$. Then, the sample size calculation is

$$n = \frac{16 \times 0.8 \times (1 - 0.8)}{0.04^2} = 1600$$

The closer $\pi$ is to one-half, the larger the required sample size will be.

## 2.6.2 Counts

When the response variable is a count, a similar procedure is followed. Here, you will be interested in the average count; we can call this $\mu$ so that $\phi = \mu$. The estimated standard error is now obtained from

$$\hat{\sigma} = \sqrt{\hat{\mu}}$$

Thus, for the desired interval to be small enough, you require that

$$n = \frac{16\hat{\mu}}{\delta^2}$$

Here, you will need to have some idea of the value of the population mean.

Suppose that you think that the mean number of children in a family is about 3 and you want an interval of 0.2. Then, the sample size calculation is

$$n = \frac{16 \times 3}{0.2^2} = 1200$$

Notice that, if you believe the mean to be 2, the sample size is smaller:

$$n = \frac{16 \times 2}{0.2^2} = 800$$

### 2.6.3 Measurements

When the response variable is a quantitative measurement, the mean will again be of interest, so that $\phi = \mu$. The method is slightly different because the standard error is not automatically given. However, the sample size formula is similar:

$$n = \frac{16\sigma^2}{\delta^2}$$

where $\sigma^2$ is the variance of the measurements. Here, you do not need to have an idea of the estimate of the mean but only of the variability. Often, this is much more difficult to obtain.

For simplicity in these examples, I have assumed that you are only interested in estimating a single value for the whole population. For the way to modify this to estimate differences among categories of an explanatory variable, see Section 3.3.4.

### 2.6.4 Complex sample designs

If you are interested in more than one response, make the sample size calculation for each and use the largest (if feasible).

If you are concerned with the precision of the estimate within each stratum of a stratified random sample, you can make the above calculations separately for each group, such as the regions of a country. In certain cases, the variability may be less within such strata than in the population as a whole, but the number in the sample will also be smaller. Generally, stratification will reduce the variability of global estimates for the whole population, especially if there are marked differences among the strata.

When you use clustering in a design, sample size calculation is much more difficult because it depends on how similar are the units within each cluster. A measure of this is usually impossible to estimate before the study begins. The precision that you will obtain in multi-stage samples is often closer to that for a sample size calculated from the sampling units than from the observational units. In an extreme example, suppose that the clusters are families and the observational units are identical twins. The amount of information available is better indicated by the number of families than by the number of children because of the similarity among twins.

Essentially, a sample size calculation requires you to assume, or to make a good guess at, what you are setting out to discover. Furthermore, you must make vast simplifications, for example ignoring all of the numerous explanatory variables that you are collecting except one crucial variable.

Sample size determination requires you to perform a delicate balancing of costs and precision. There is rarely any point in collecting data on a sample that is so large that it provides much greater precision than that actually needed. But you may have to abandon any hope of high precision if the cost of a sufficiently large sample is too great.

## 2.7 Summary

Samples are often to be preferred to censuses because they can provide more accurate information at less cost. Random sampling is necessary in order to avoid unknown biases and to yield a measure of precision of the estimates.

The observational unit, about which you will be collecting information, may differ from the sampling unit, chosen at random in the population. The sampling units in the population are defined by the sampling frame.

A pilot study is almost always necessary to test the field procedures and to obtain an idea of the variability in the population under study.

You may use many different types of instruments in sample surveys. Those most useful for subsequent statistical analysis include observational schedules, tests of knowledge, and questionnaires. You should take great care in their construction so as to obtain accurate information. You may administer them directly, say by interviews, or indirectly, for example through the post. Carefully choose the field investigators and adequately train them for the specific procedures that you will use.

Sampling error arises from biases and from the random variability in the sample. The biases may result from inadequacies in the sampling procedure or from missing responses, as well as from problems with the instruments or investigators.

The main sampling designs are simple random samples, stratification by randomly choosing fixed proportions of individuals within subgroups of the population, and clustering by randomly choosing whole groups of individuals. On the other hand, case–control designs are sometimes useful when you are studying some rare event. For changes over time, you will need to do repeated sampling, as in a panel study.

Sample size calculations will provide you with a rough idea of how many individuals you must observe in order to obtain a desired degree of precision or to detect some effect of interest.

# 3

# Experimental trials

## 3.1 Basic principles

The main feature that distinguishes an experimental trial from a sample survey is that you perform an intervention or *treatment* instead of simply observing things as they are. Fundamental advantages of the experimental method are that:

- causality can be empirically studied;
- a complex causal problem can be attacked by proceeding in a series of simple steps.

Thus, you can break a problem up into simple questions to be explored by separate trials with simple causal assumptions. However, as we have seen, the major drawback, especially with human subjects, is that they cannot be chosen randomly, making generalization to a larger population difficult.

In a trial, the principal sources of exposure will be under your control; you decide to whom each is applied. The smallest entity that might have received a different (sequence of) intervention(s), when they were allocated in the study, is called the *experimental unit*. As with sampling units in a survey, there may occasionally be several levels of experimental units, such as classrooms and children.

In certain scientific investigations, you will not precisely know the *causal factors* so that a goal of the study is to determine which are relevant. In other situations, you may possibly apply several distinct types of intervention in various combinations. In either case, it may often be desirable for you to include several causal factors simultaneously in the same trial.

You will often begin your investigation of any relatively complicated phenomenon with a general survey of the effect of a variety of changes on the system. Next, you may test more closely your ideas about how parts of the system really work. Most often, you will need a series of trials, with your initial ideas being corrected at each step.

Even trials with direct practical political or commercial aims may allow you to include special treatments yielding fundamental knowledge about the process under study. A good check on reliability of a trial is agreement with previously established results in the field. Often, it is worth including some specific 'standardized' treatment solely for this objective.

The basic requirements for you to perform an acceptable experimental trial include:

- as simple, but efficient, a design as possible;
- freedom from systematic error or bias;
- sufficiently precise and exact measurement of the response;
- a measure of precision of the results;
- wide validity of the results.

Various procedures are necessary to ensure these goals. Three of the important steps are choice of:

1. treatments;
2. experimental units;
3. types of measurements observed.

Generally, the choice of treatments, except perhaps for the form of control, is a technical question specific to the subject under study. Once you have settled these three questions, you can elaborate the overall design of the trial.

### 3.1.1 Controls and placebo

Experimental trials can only allow you to identify the effect of an intervention by comparison with something else. Observation of past conditions, before intervention, called historical controls, is not sufficient because there can be external evolution over time or unrecorded differences among the subjects involved. Thus, you must include a simultaneous *control* treatment for a trial to be worthwhile. In this way, you construct experimentally the causal factor under study.

In many trials, you must randomly assign subjects either to treatment or to control for valid comparisons to be feasible. Inform the participants that they will be (blindly) randomized either to the control or to the new or active treatment. That is, where possible, they will not know which they receive. The protocol should be available for inspection by all participants.

Cross-over trials (Section 3.3.1) are somewhat different in that all subjects will receive the treatment(s) and control, but in different orders.

A *placebo* control is an inert treatment that appears, in all external aspects, to be identical to the active treatment. It reduces the chance of subjects guessing which treatment they are receiving. Thus, for example, in testing drugs, the placebo would be an inert substance, identical in taste, appearance, smell, density, and so on, to the active drug.

Using a placebo will allow you to distinguish the true side effects of the intervention. If your control involves no intervention at all, this can create problems of interpretation. Those receiving the intervention may be reacting simply because they are getting special attention (the placebo or Hawthorne effect), and not because of the specific nature of the treatment. Thus, it is preferable if members of all groups believe that they are receiving equivalent special treatment.

The role of a placebo is not to deceive the subjects into thinking that they are receiving the active treatment, but to leave them in doubt as to which treatment they are receiving. Thus, ethically, a placebo is only possible when randomization of treatments is used.

If several active treatments, dissimilar to each other, are involved in a trial, several corresponding placebos may also be necessary. This can pose the problem of overburdening the subject. Sometimes, it may also be useful to include two control groups, with and without placebo.

In many situations, it is ethically impossible not to supply an active treatment. Then, the new treatment is generally compared to the existing or standard treatment as the control, instead of to a placebo.

Always remember that a treatment effect, between two experimental groups (of volunteers), that is perfectly genuine under trial conditions may be quite different when introduced on a large scale in realistic conditions.

### 3.1.2 Choice of subjects

As for sample surveys, carefully plan what is to be the eligible population. Specify eligibility criteria in advance, in the protocol. You must not determine them after selecting subjects and assigning them to an experimental group.

The fact that an intervention is involved means that it is generally impossible to choose a random sample of subjects to participate in a trial. You cannot impose the intervention upon them; they must voluntarily agree. Thus, subjects in such a trial are never representative of a larger population. Even if there are no refusals among those in the eligible population whom you ask, this does not make the subjects representative of some larger population.

Eligibility criteria may render the chosen subjects even more unrepresentative. In a medical study, if subjects are required not to be taking any other medication, they will be healthier than average. The more restrictive are the exclusion criteria, the less generalizable will be the results. Extrapolation is always a risky process.

At the same time, the subjects should not be unrepresentative. Your primary scientific goal in an experimental trial is to investigate what effect some intervention can have by showing what effect it had in a particular case. In other words, the test intervention should not be applied to subjects who would not receive it under normal conditions after testing is completed. (A exception might be the initial testing of a drug for tolerability on healthy volunteers.)

Often, your choice of investigators may conflict with your choice of subjects. Centres of excellence, for example hospitals or schools with highly trained research staff, will generally provide select and unrepresentative subjects.

Your next step will be to define what will be the experimental units. Factors to take into account include:

- their size, for example, classroom or child;
- how representative they need be;
- how close to realistic are the conditions in which they will be studied;
- whether responses on the same unit can be observed several times, perhaps under different treatments.

In most trials, the size of the units will not be in question, because it will be the individual subject.

### *3.1.3 Randomization of treatment*

In a trial, you want to compare a set of two or more treatments, including the control, with a group of subjects assigned to each. These groups should initially be as alike as possible in all ways so that, in the subsequent analysis, you will only need to compare them on the characteristics of direct interest. Individuals need not be equivalent; you are studying the group reaction.

The only objective way for you to ensure this similarity is by randomization. If the choice of treatment is made by the investigator, or by the subject, there is great freedom for unconscious or intentional bias. Thus, even although you generally cannot choose subjects at random from some larger population, use randomization for the allocation of treatments among the participating subjects.

The equivalence achieved by random assignment is probabilistic. It is not inevitable that a correctly implemented randomization procedure will result in similar groups, although the larger the groups the more chance of their not differing much. A successfully implemented random assignment procedure does not guarantee that the initial comparability among groups will be maintained over the course of a trial. The major risk is differential attrition related to treatments. In many trials, this can be an important object of study in its own right.

Thus, this randomization does not rule out all threats to causal validity.

- Subjects in a control group may imitate those under a treatment.
- If a control group is felt to be unjustly treated, compensatory measures may be available to it from elsewhere.
- A group that feels it is receiving less desirable treatment may exert special effort in compensation or it may become demoralized.

You may prevent some of these by blinding where this is possible, as discussed below.

If certain combinations of treatment allocation, that might arise randomly, are undesirable, these can be specified in advance and eliminated. For example, when subjects are admitted to a trial sequentially, the first half of subjects might all randomly be assigned to receive one treatment and the second half the other. Such possibilities are rare and generally only occur in very small trials.

Subjects meeting the selection criteria are said to *enter* the trial. Any exclusion after randomization, for whatever reason, even discovery that eligibility criteria have not actually been met, may upset the randomization balance. Before entry, every subject must be regarded as suitable for any of the treatments under study. If this is not true, equivalent groups cannot be constructed and comparisons will be impossible.

Your treatment allocation system should be such that the people entering subjects into the trial do not know in advance what is the next treatment to be assigned. If they did know, they might decide that the next treatment was inappropriate for the subject and not enter that person into the trial. Thus, randomization is important as a measure of concealment of the order of assigning treatments, because any system in the sequence might be detected.

If the same investigator cannot administer the treatments to all subjects, design the study so that investigators are also distributed so as to give all treatments in some random fashion. In this way, you can avoid confounding investigator bias with treatment.

In simple randomization, you generate a series of random numbers by computer. If only two treatments are involved, you can assign the treatment of each subject sequentially, with odd digits in the series indicating one treatment and even digits the other. With three treatments, 1, 2, and 3 might indicate the first, 4, 5, and 6 the second, and 7, 8, and 9 the third, zeros being ignored. You can easily adapt such a procedure to weighted randomization, with different proportions in the groups.

Start treatment as soon as possible after entry and randomization of each subject, to avoid any intervening changes in the state of the subjects. If at all possible, the waiting time should be equal in all groups to avoid differential losses. In many studies, entry must be staggered in time, as subjects become available.

If it is necessary to ensure that exactly the same number of subjects is assigned to each group, you will require somewhat more complex schemes. You can also use stratification and cluster methods, as in sample surveys (Sections 2.5.2 and 2.5.3). Here, the first are often called *blocks* and *classification factors* and the second *plots*, both to be distinguished from the causal factors of interest. As in multi-stage surveys, there may be several levels of 'plots', and several sizes of experimental units, for example in a *split plot* design.

### Cluster randomization

For intervention studies involving, say, new disease prevention procedures or educational methods, randomization of clusters is often more appropriate than that of individuals. This may be necessary for various reasons.

- The intervention may have to be at the group level, as in community health services or an educational curriculum.
- If allocation is at the individual level, control subjects may benefit from observing or communicating with the treatment subjects.
- You can capture the mass effect of interaction among individuals in a more realistic setting within each cluster.

- In a context where transmission is possible, such as an infectious disease, you can study both susceptibility and infectiousness within each cluster.
- In a large study, it may be infeasible to allocate treatment and control to different individuals in the same group even although it would be possible within a small group.

The clusters might be villages, schools, medical practices, factories, and so on. In many situations, cluster randomization provides experimental information that is closer to that which will prevail if the intervention is subsequently extended to the whole population.

As with cluster sampling, there is a trade-off between a few large clusters and many small ones. The former are cheaper but may also reduce contamination between treatment and control. On the other hand, the latter allow you to control variability among clusters so that the averaging effect of randomization can work. If you must use large clusters, you may randomly sample individuals for further follow-up within each cluster. If clusters are geographical, you may concentrate sampling near the centre to avoid contamination from neighbouring clusters.

### 3.1.4 Blinding

Even unconsciously, observers' judgements can be affected by knowing which treatment each subject is receiving, or by knowledge of previous measurements on that subject. In the same way, a subject knowing that he or she is receiving active treatment may react differently than if the same subject knew he or she had been assigned to the control group.

*Blinding*, or *masking*, refers to the fact that the people involved in a trial are not aware of the treatment a subject is receiving. In a single-blind trial, the subjects are not aware, whereas, in a double-blind trial, neither the investigators directly involved nor the subjects know. The latter is preferable but is obviously impossible in many contexts. If you cannot blind the investigators directly involved, then you may be able to use external assessment of responses.

A double-blind trial has the enormous advantage that the investigators cannot, even unconsciously, bias the results in one group and can objectively evaluate progress in all groups. Indeed, in many cases, it will be advantageous to use a triple-blind trial, whereby the statistician conducting the analysis is also not aware of which group corresponds to which treatment.

For blinding to be feasible, a placebo treatment must be available. As with the use of a placebo, the subjects should be informed of the protocol; blinding is only ethical under randomization of treatments.

### 3.1.5 Primary end-point

Carefully define the main response variable that will be observed, as well as when this will occur, called the *primary end-point*. Describe in detail the way in which you will measure this and closely control it. However, this is usually a technical question, specific to the subject matter being studied, and not a statistical question like the questionnaire construction in sample surveys. Nevertheless, you should take similar care about recording the results and ensuring that they will be susceptible to statistical analysis. Thus, it will be useful if you review Section 2.3.

Generally, you must follow the effects of intervention over a certain period of time. In other words, you will need to monitor subjects to detect some change. Take into account:

- the resources available;
- the frequency of visits required for normal monitoring of the process;
- the inconvenience to subjects of frequent evaluation;
- the number of measurements required to provide an adequate comparison among treatments.

Often, you may want to focus on evaluation of responses at the beginning and end of some fixed period.

When you are performing interventions, the observations that you measure on subjects, in addition to the actual treatment(s), generally can take five main forms:

1. the *baseline* response before treatment begins;
2. supplementary baseline explanatory variables and, as well, sometimes time-varying ones;
3. the principal response end-point of interest;
4. secondary responses, such as side effects;
5. monitoring, including compliance.

In many situations, you may find it useful to make preliminary observations on the subjects before treatments begin, called baseline response values. Several recordings may be necessary to avoid random fluctuations. These will provide indications of the variability among the subjects, and the reliability of measurement, that may be useful in controlling and interpreting the results. Indeed, if you are measuring change or improvement under treatment, recording the baseline response is essential.

Especially if there is substantial variation among subjects, it may prove useful to plan to collect concomitant observations on other variables besides the baseline response. The main condition for these to be useful in the subsequent analysis is that they not be affected by the treatment. You can easily ensure this by taking the measurements before treatment begins, but this is not always possible. If you believe that they influence the subject's response to treatment, you may use them as prognostic factors.

Take particular care with subject identification: there will usually be several forms to be filled out at different points in time that must later be linked together. Investigators must have the appropriate forms available at the right time. Send out requests for them to be returned promptly.

Note and record *all* reactions of all groups equally. A checklist of possible side effects and the reasons for non-compliance may be useful, although this has the same dangers as closed questions in surveys (Section 2.3.2).

If judgements or interpretations of the data recorded for each subject are required, make them before disclosure of which group had what treatment.

For many types of trials, it is important to conduct some form of follow-up study to determine what are the long-term effects of the intervention, perhaps years after the treatment has ended.

### 3.1.6 Missing values and non-compliance

Because your subjects are volunteers, the problem of refusals found in sample surveys generally does not occur here. On the other hand, missing values will occur for certain observations. Where applicable, pay special attention to provision for treatment during holidays and weekends.

In addition, because of the lengthy time period often involved, the number of drop-outs can be considerable. Sometimes this will be unavoidable, as with deaths, people moving away, and so on. However, in other cases, treatment may be stopped or changed because of side effects, and so on. (Intentional drop-out might be considered to be a form of refusal.) If drop-outs are not linked with the intervention, they create no fundamental problem (except loss of information) for internal validity, although problems may arise for external validity.

You can make several types of checks on differential attrition:

- Is the rate of drop-out the same in all groups?
- Are the reasons for dropping out the same in all groups?
- Are the pre-randomization covariates still comparable for those remaining in the groups?
- If baseline response measures are available, are they still comparable for those remaining in the groups?

A problem that does not arise with surveys is *compliance*: do the subjects actually follow the treatment assigned to them? This can be very difficult to judge, because, for example, patients can discard pills instead of consuming them and teachers can go back to their old teaching methods. In some cases, you can assess the probability of adherence to the assigned treatment and use it as a screening criterion for entry to a trial. Once the trial begins, use all realistic procedures to ensure the maximum compliance possible. Record measures of the extent of non-compliance. The need for screening and compliance enforcement will usually make it more difficult for you to generalize the results to applications in more realistic situations.

If your goal is to study the effect of intervention in realistic conditions, compliance may not be a problem because the same thing could be expected to occur under normal conditions for the population as a whole. Then, statistical analysis is carried out on *intention to treat*, that is, on the treatment assigned to a subject, not necessarily that actually followed.

Related to compliance is the wider issue as to whether subjects are reacting in a reasonably normal way in the experimental conditions so that fairly general conclusions can be drawn. You can take certain steps in an attempt to ensure this:

- Check the reactions on pilot subjects before beginning the study.
- Redesign the study to reduce suspicion, if necessary.
- Provide clear instructions.
- Emphasize personal anonymity.
- Minimize aspects that appear to test the subject.
- Use non-invasive instruments to obtain response measures.
- Separate the experimental manipulation and the response measurements as far as possible.

- Develop special instruments to detect if subjects are simply trying to please the investigator.

## 3.2 Ethical issues

Because you are imposing an intervention, ethical issues are much more critical than in a sample survey. However, many of the basic questions are the same; your should review Sections 1.1.3 and 2.2.3.

### *3.2.1 Ethics committees*

For experimental trials, the conflict between present individual and future collective ethics may be particularly evident. Each subject should receive the most appropriate treatment now, but, in the future, all subjects may collectively benefit from an intervention that has been shown to be superior. Weighing the merits of the two for a given trial is often a delicate task. Some judgement must be made as to whether the impingement on the individual is such, as compared with the possible benefit to society, that the trial should not take place.

For most areas of study, when interventions using human subjects are involved, you must obtain permission from an ethics committee. Generally, you must submit a full protocol describing the study, with ample justification that the objectives are worthwhile, that the design is efficient, and that the rights of the subjects will be protected.

### 3.2.2 Informed consent

The main question at issue is whether it is ethical to withhold a treatment that might perhaps give benefit. On the one hand, the value of the intervention is not proven or there would be no need for the trial. On the other, there must be some basis for considering the new treatment, or a trial would not be undertaken. In no case should the control group be disadvantaged by participation as compared to their non-involvement in the study.

The decision often depends on the gravity of the condition being treated, it being impossible, for example, to withhold treatment in a life-and-death situation (which treatment, the old or the new?). But it may be unethical to introduce a new treatment into general use, if it has been poorly or inadequately tested. All risks do not lie on one side; what is new is not always best.

On the research side, ideally no investigator should participate who believes that one treatment is clearly superior. An investigator should not enter a subject in a trial, if he or she believes a particular treatment to be preferable for that person. In other words, you are conducting the trial because no one knows which treatment is better.

Unrandomized trials are almost invariably unethical because they involve subjects with risks when the results of the study will be unreliable. The same might be said of any trial that does not use an optimally efficient design.

Questions that you must face include:

- Should you obtain the subjects' informed consent?
- Is the new treatment safe and unlikely to bring harm?

- In a trial with inert control, can you ethically withhold treatment?
- Is it acceptable to use a placebo treatment?
- What type of subjects can you acceptably allocate randomly between different treatments?
- Is it all right to use double-blinding?

(See Hill and Hill, 1991, pp. 212–214.) For most questions except the first, there is rarely an unequivocal answer.

Inform potential subjects in detail about the conduct of the trial, including the alternative treatments that will be involved. You must guarantee confidentiality. Potential subjects can decline to participate. If they agree to take part, they generally will have to sign a form stating that they understand the trial, called *informed consent*. The question may arise as to whether they actually do understand or not. *Legal requirements as to such consent vary greatly among countries.*

If the experimental unit is a group, then you must generally obtain special permission from the leaders of that group, such as school directors or community officials. Although you should inform the members of the group, obtaining active individual consent is not always possible. However, if individuals are to be followed up and tested, you will require their consent.

### 3.2.3 Interim analysis

One of the main reasons for monitoring a trial is the ethical concern to reduce the chance of subjects receiving a treatment known to be inferior. Thus, when subjects are to receive treatment over a long period of time or if their entry to the trial is staggered, interim assessment of treatment differences, where possible, is essential to make a trial ethically acceptable. Once you can reach a conclusion, you will enter no new subjects, and you may perhaps change those in the trial to a more appropriate treatment.

Generally, you will check the primary end-point response, usually one main treatment comparison. Specify in the protocol the formal 'stopping rule' for the trial, that is, the criterion indicating sufficient superiority of a treatment so that the trial can be stopped. For this to be possible:

- the time lag between subject entry and meaningful response measurements must not be excessive, especially in comparison to the total time during which individuals will be entered;
- data entry must be kept up to date, with *all* forms quickly returned, not just an unrepresentative set, such as the optimistic ones.

Interim analyses should be relatively simple, sufficient only to determine if the trial is to continue or not.

You will generally have interim analyses performed periodically and not continuously, with the frequency specified in the protocol. They should be performed by people not directly involved in the investigation, and the decision on continuing made by the committee responsible for the trial. The decision to stop a trial will never be purely statistical.

The results, if the trial is to continue, should remain confidential so that subsequent recruitment and responses will not be biased. They can influence both subjects and investigators.

## 3.3 Designs

In most trials, only a limited increase in precision can be obtained by modifying the instruments used. Trials under very controlled conditions usually cease to be representative of practical conditions. Precision will depend much more on intrinsic human variability and experimental design, including sample size.

Take care not to use too many different treatments because the ability to detect treatment differences depends primarily on the number of subjects per treatment, not on the total number in the trial.

The two principal designs used in experimental trials with human beings are the parallel and cross-over designs. The scientific ideal would always be to administer all treatments to all subjects, because different subjects may react in different ways to the various treatments. However, this is rarely possible for various practical and ethical reasons.

### *3.3.1 Cross-over designs*

In certain special circumstances, it may be possible for you to apply several treatments sequentially to each subject. If you can apply these in different orders, you require a *cross-over design*: you randomly assign each subject to receive some given sequence of different treatments in successive periods. Often, you can use some form of *Latin square design* to balance the numbers in the different sequences: each treatment appears once in each period and in each sequence.

Note that this is quite different from applying all treatments to all subjects, but always in the same order. Then, treatment effects cannot be distinguished from a temporal change.

Obviously, you can only apply such designs in situations where:

- the condition under study is chronic, with no trend in time;
- the measured response is recurrent;
- the condition is not fundamentally modified by a treatment.

These designs are really the only possibility when subjects may differ in their responses to the various treatments, that is, when there is the possibility of a subject–treatment interaction. They are more useful for one-shot treatments, than long-term ones.

Cross-over designs have the major advantage that you can compare the different treatments on the same subjects. This generally means that the sample size can be smaller for the same precision.

The major disadvantages are as follows:

- The effects of earlier treatments may *carry over* to modify responses under later treatments.
- Drop-out may occur between treatments so that no within-subject comparison is possible.
- Systematic differences may appear between periods, perhaps simply due to the learning effect of being treated in the first period.

To counter carryover, either plan an adequate *wash-out period* between the end of one treatment and the beginning of the next or allow a 'burn-in' period at the beginning of each treatment, during which you do not measure (or, at least, do not analyse) the response values. In certain contexts, the first possibility will present ethical problems, if you have to withhold treatment completely during that period.

If period effects are present, variability among individuals has been replaced by variability over time within individuals.

Many different types of specific cross-over designs exist. These depend on the number of different treatments and periods, on whether each subject receives each treatment in turn, and so on.

### 3.3.2 Matching

If a cross-over trial is not possible, it may still be possible for you to find pairs of subjects with closely similar characteristics and to allocate treatment and control randomly to the two members of the *matched pair*. The major problem here is to have a large enough pool of subjects from which you can draw ones with the appropriate characteristics simultaneously. If matching is made too precise, it will be difficult to obtain two suitable subjects at the same time.

One advantage of this design is that it may permit you to detect what kinds of subjects are most susceptible to benefit from each treatment. For example, for the majority of subjects, response differences to treatment may be negligible but, for a minority, one treatment may be vital.

### 3.3.3 Parallel designs

In a *parallel design*, you randomize subjects to different treatments; they then stay on them for the whole trial. Most such trials involve a new treatment and either a control or standard treatment. Randomization to the treatments allows you to make the assumption that the groups are comparable in all relevant aspects. This is the standard and most common type of trial.

### Factorial designs

Treatments should differ qualitatively in single, specifically identifiable, ways so that your interpretation of response differences will be clear, with unique explanations. In other words, more complex treatments should be split up into several simple factors. Remember that an effect generally has several causes. In addition, it may not act the same way in all situations.

If you are to study several types of intervention, the factors, it is almost always preferable to use them simultaneously in a *factorial design* that includes all possible combinations. The different treatments within a factor are often called *levels*. Examples of pairs of treatments would include new textbooks and new teaching practices, or two complementary types of medication. Thus, in the simplest case, for example, with two such treatments, each at two levels (control and active), you would have four groups: (1) control only (perhaps two placebos); (2) first type of treatment only, perhaps with a placebo; (3) second type of treatment only, perhaps with a placebo; and (4) both types of treatment. Factorial experiments may also include combinations of treatment and classification factors.

Factorial designs have several advantages: you can

- obtain more information from a smaller number of sub-

jects, that is, higher precision at lower cost;
- study the *interaction* among types of treatment with respect to the response;
- perhaps extend the validity of your conclusions by the inclusion of a classification factor that increases the variability of the conditions of application;
- check on whether treatments might have different effects under different conditions, a causal factor interacting with a classification factor.

Two factors are said to interact, in the statistical sense, if the effect of one factor on the response changes depending on what value the other has. If there is no interaction, the difference in response between two levels of one factor is the same no matter what level the other has. Thus, for the statistician, the term 'interaction' does not have the usual common-sense meaning.

These considerations for factorial designs do not imply that you should plan the biggest and most complex trial possible.

- Take care, especially at the beginning of an investigation, not to commit yourself to one big trial; small preliminary trials may indicate the proper line of attack.
- For a real understanding of many problems, a series of small trials is more appropriate than one large trial, each one designed in the light of previous results.
- Large and complex trials are difficult to organize.

In most cases, you should consider no more than two or three factors.

Factor levels may not necessarily be qualitatively distinct, especially in factorial designs. We may distinguish:

- fixed qualitative factors, each level being of intrinsic interest;
- ranked categories, such as slight, moderate, and severe;
- quantitative factors, with levels fixed at certain arbitrary levels, such as doses of a drug;
- factor levels that are assumed to be 'representative' of some larger population of possible levels, although rarely chosen at random.

With quantitative factors, you can study changes as a *response curve* (or *surface* if in several dimensions). Because only a few levels are used, analysis will generally require you to make the assumption of some smooth functional statistical model connecting them together.

## Sequential trials

When it is imperative not to continue an inferior treatment (once this becomes known), you may use a (group) *sequential trial.* You continue the study only until it is clear that one treatment is superior to another. To do this, you must analyse the data after the results of each subject or group of subjects become available. For this to be possible, subjects must enter the trial sequentially and results must be available relatively quickly after administering the intervention.

Such trials have the additional advantage that, on average, you will require a substantially smaller number of subjects than in a fixed sample size trial, particularly if there is a big difference in response between the treatments. However, when designing such a trial, you should realize that the conventional probability levels for significance tests and confidence intervals are much more difficult to calculate than when the size is fixed in advance.

### Equivalence trials

Trials to determine if a new treatment is equivalent to the existing standard one are particularly difficult to design because of the danger of detecting no difference simply because the study was too small. Those judging the results of such a trial must be convinced that it was properly conducted.

## *3.3.4 Sample size*

Sample size calculations are very similar to those for sample surveys. You should first (re)read Section 2.6. Maintaining a trial at its minimum size necessary to provide the required precision is here even more imperative than for surveys because of the risks involved in imposing interventions. The sequential trials mentioned above are one means to this end. If you cannot use one, you will need to calculate the minimal sample size.

The *power* of a trial is its ability to detect a difference of interest, due to an intervention, if it really exists. This is primarily a function of sample size. Too small a trial is a waste of resources and exposes subjects to useless risks. In equivalence trials, too small a sample will ensure that treatments appear to be equivalent!

Consider the simple case of two treatments, say active and control, and equal numbers in each group. The sample size is calculated for differences in response at the primary end-point.

The first step in a sample size calculation is to specify the smallest difference in response among treatments, $\delta$, that would be of importance. This is not a statistical, but a scientific question. Then, you can apply the formula of Section 2.6, but where you replace the 16 by 64 to obtain precision equivalent to the two standard errors used there. In other words, you multiply by four all values that you would have calculated for the examples without covariates given above for sample surveys.

As an example, for binary responses, the value of $\pi$ will be the expected average probability under the two treatments. Suppose that you want to detect a difference in probability of response of 10%, between say 75% under control and 85% under active treatment. The average is 80% and the sample size is calculated to be

$$n = \frac{64 \times 0.8 \times (1 - 0.8)}{0.1^2} = 1024$$

that is, 512 in each treatment group. You can use similar procedures for counts and measurements, as in Section 2.6.

In more complex trials, one particular contrast between two treatments may be of particular interest and you might use this to calculate the required sample size. Otherwise, you will require more complex techniques. In a sequential trial, the standard error calculated at a first stage will generally give you a good idea of how many subjects are still required to attain the desired precision.

It is particularly important, when you begin a trial, to ensure that you foresee a large enough sample size, that sufficient subjects will be available to fulfil that goal, and that your finance support is adequate. Do not undertake a trial that will result in too few subjects to meet scientific requirements of precision.

When subjects enter into a trial sequentially, as they become available, you must estimate the *accrual rate* accurately in order to ensure that enough people will be available in the time planned for the trial. This will often be an overestimate mainly because of:

- over-enthusiasm of the research workers planning the trial;
- ineligibility of some people;
- refusals;
- loss of interest over time if the study is too long.

If you cannot meet the sample size in the allotted time, you will need to:

- increase the accrual rate, for example by using more centres or changing eligibility criteria;
- reduce precision;
- increase the time period; or
- stop the trial, if it will not be possible to detect relevant differences among treatments.

Many trials that find no evidence of treatment differences are too small to reach a reliable conclusion.

## 3.4 Organization

### 3.4.1 Clinical trials

In medical research, experiments are usually *clinical trials* to evaluate the effectiveness of some treatment(s), such as drugs, surgery, physiotherapy, diet, health education, and so on. These trials thus involve patients with a given medical condition and are designed to provide information about the most appropriate treatment for future patients having the same condition. The vast majority of clinical trials are concerned with evaluating some specific drug, most often conducted, or at least financed, by a pharmaceutical company.

A complete study for a new therapy requires evaluation of *safety*, *efficacy*, and *quality of life*. Especially for a drug that is to be commercialized, preclinical research begins with animal experiments testing for safety. Then, human experimentation can generally be classified into four phases:

1. *Phase I*: initial study of pharmacology and toxicity, usually with healthy subjects, to determine safety at various doses, including side effects;
2. *Phase II*: small-scale, often non-comparative, clinical investigation on patients, to screen out ineffective drugs and to determine dose and other characteristics of the therapy;
3. *Phase III*: full-scale evaluation of an apparently effective treatment in its final form, as compared to a control or to standard treatment;
4. *Phase IV*: surveillance after approval for commercialization, monitoring long-term adverse effects.

The first two phases involve tightly controlled scientific investigation of medical aspects, whereas the third is closer to realistic administration of the therapy once commercialized. The first two are exploratory, providing hypotheses that can be tested with the results of the third. The fourth may often take the form of a sample survey, rather than a trial.

Phase II trials were traditionally often uncontrolled and not blinded; this could bias the results, for example by the enthusiasm of the investigators. Many early studies of this type have suggested that the new treatment is highly effective, only for this apparent benefit to disappear when more carefully tested in Phase III. Thus the use of randomization is increasing in Phase II studies.

### 3.4.2 Multi-centre trials

An experimental trial may be organized in one centre, a hospital or school for example, or, more often, in a number of centres. The latter will create an effect of clustering, as in sample surveys. The advantage of a *multi-centre trial*, over using just one centre, is that it includes much more variability among subjects so that there is more chance of the results being generally applicable. This should be true even although the centres recruited will not be representative of centres in general. Centres are chosen for reasons of cost, efficiency, convenience, the research group's reputation, and so on, but rarely because they are typical.

The centres involved will usually be rather variable, for size, geographical, subject recruitment, and other reasons. It will often be preferable to allow these variations to enter the study, particularly if they are commonly met in practice, rather than to maintain too high a degree of standardization, difficult to accept either during the trial or subsequently. If different centres provide different results, then generalizability is in question. This is important information to obtain.

Often, the choice will be between a large number of subjects rapidly recruited in a number of centres with relatively short follow-up and fewer subjects with a long follow-up. The former design can present major administrative problems, including ensuring the constant use of the same procedures in all centres. The latter may suffer from changing investigators, loss of interest, and so on.

The difficulties that may arise with multi-centre trials include:

- complex planning and administration;
- excessive expense;
- demotivation of investigators;
- non-uniform eligibility requirements for entry;
- problems with monitoring;
- greater risks of missing data, for example from lost forms;
- deviations from the protocol, including problems of non-compliance;
- lack of uniform quality control of data.

Carefully weigh these factors before undertaking such a trial.

### 3.4.3 Longitudinal trials

Most often, you must follow subjects over a sufficiently long period for the treatment to have an effect. In this sense, all trials are longitudinal. However, in many trials, a specific end-point is defined: the final response of interest. Because this occurs at a given point in time, the study is fundamentally cross-sectional.

In other cases, you will measure responses, at fixed or convenient times, over a considerable period. Thus, you will have a sequence of response values for each subject. This is a true longitudinal study. Notice, however, that randomization to treatment only assures comparability among groups at the moment of allocation. As the subjects evolve over time, their characteristics will be modified, depending on their previous history, and hence may no longer be comparable.

## 3.5 Summary

An experimental trial is an intervention involving human beings. The major advantage of such a trial is that you can empirically study causality. However, because you must use consenting volunteers, the results will often not be easily generalizable to more realistic situations outside of the controlled context of the trial.

You can only identify the effect of a new treatment by comparing it with something else, usually a placebo or the standard treatment. The only objective way to ensure that the groups receiving the different treatments (including placebo) are similar is to assign subjects randomly to them. To avoid biases in reactions of both subjects and observers, blind the trial whenever possible, so that those directly involved do not know who has what treatment.

The primary end-point is the response of main interest to be observed at some well-defined time. Baseline measurements will also usually be useful.

Missing values, including drop-outs, will be of concern. One problem specific to trials is compliance: are the subjects actually taking the treatment assigned to them?

Because you are applying an intervention, ethical issues are crucial. All subjects must give informed consent. You must use procedures, such as interim analysis, to stop the trial as soon as possible if it is clear that one treatment (including the placebo) is superior so that no subjects continue longer than necessary on an inferior one.

Cross-over and parallel designs are the principal ones used. The former have the advantage that the various treatments are compared on the same people, but this is not possible in many cases.

In medical research, clinical trials have a central place in the development of new procedures. Both safety and efficacy must be determined in a series of phases from initial exploration to final confirmation and long-term surveillance.

Both multi-centre and longitudinal trials have their own special complications.

# 4

# Data analysis

## 4.1 Data handling

Efficient data management is essential to minimize errors. This usually means that you should employ specialized staff, neither the investigators nor the statisticians, for this task. Your basic goals will be to have data that are:

- complete;
- accurate;
- uniform or properly standardized;
- coherent.

In many ways, achieving these goals will provide you with a check on the actual data collection process.

### *4.1.1 Data entry*

Once you terminate data collection, your first step will be to check that all forms have been returned. Also visually check them for any obvious errors and omissions, and seek the necessary rectifications from the field investigators. Then, have the data transferred to magnetic form for analysis by computer. The main steps are:

1. coding the results in a suitable form for computer treatment (you may have incorporated this stage into the original questionnaire);
2. typing the coded values into the computer (in certain cases, this may also be possible electronically, as with bar codes and light pens);
3. editing the resulting file(s) to remove errors.

This can be one of the most costly and time-consuming operations in a study. It is the one where errors are quite possible, and where you can most easily avoid them, with proper care.

The main sources of error are:

- recording, as when an instrument is misread;
- deliberate, due to falsification either by the investigator or by the respondent;
- transcription when observations are copied;
- typing (during data entry).

You should have the data entered using some *database management system*; some of the major statistical packages contain one or you can use a general-purpose one. If the mass of data is of a reasonable size, you should have an entry grid developed within the database system; this will show the coder, at all times, what variable is currently being entered and prevent entry of impossible values. Thus, if a question has a binary response, such as yes/no, that is to be coded as 1 or 2 (plus a missing value code), no other values will be accepted. This removes one of the most common sources of fundamental errors in the database.

If you are using several forms for each observational unit, whether from different levels of clustering or longitudinally over time, such a database system will also provide a convenient means of linking them together. Complications may arise, especially in experimental trials and longitudinal studies, because:

- subjects have unequal amounts of data recorded;
- data on each subject accumulate over time.

In trials with staggered entry, earlier subjects will have more data available than later ones at any particular moment. Carefully choose ways of handling these problems.

You can reduce errors if you use the same code for missing values in all variables, although this may not always be possible. Some computer programs have a special code symbol for this, but, if you use it, no other software may be able to read the file. Most often the code chosen is a set of nines, the number of digits corresponding to the maximum number of digits occupied by the variable. The use of blanks to indicate missing values is especially dangerous because much statistical software cannot distinguish these from zero values.

As explained in previous chapters, plan the ways in which each variable will be coded for entry into the computer at the stage of preparing the instruments for data collection. You should have planned data collection in such a way that no intermediate calculations are necessary until after the data are in the computer. The basic types of coding are:

- counts;
- measurements, for which the units should be clearly specified;

- discrete categories, for which special coding will be required.

Let us consider each of these in turn.

**Counts** These take integral values that cannot be negative. There may be some upper limit to the reasonable values that they can take.

**Measurements** Almost all measurements are positive values. Common errors involve:

- a misplaced decimal point;
- digit preference, whereby observers tend to round to say zero or five for the last digit of each number;
- confusion about the type of units – in an international study, feet or yards for lengths in one country and metres in another;
- mistakes in the size of the units – some values in millimetres and others in centimetres, or some in days and others in weeks.

The last can be difficult to detect, especially if the units are not mentioned on the data form.

In international studies, extreme care must be taken with numerical dates; not all countries use the day–month–year order. If years are only recorded to two digits, errors can easily occur at the turn of the century.

**Discrete categories** Many statistical packages allow you to code discrete categories by alphabetic or numerical means, but some require all values to be numeric. Use of alphabetic codes may restrict future analysis to the former packages, but may reduce the risk of errors at data entry.

More general errors include:

- transposition of digits;
- repetition of the same value in two successive rows or columns, perhaps with the effect of displacing all subsequent values in that row or column;
- hidden time effects, whereby the value observed depends, for example, on the time of day or the time of year.

The safest, but most costly, way for you to ensure that data are entered correctly is to do it twice with different people. Then, a computer program can check for differences between the two files. Any found must obviously be corrected! This process, however, will only detect errors in the transfer of data from the forms to the computer. It will not find errors in originally filling out the forms.

It is usually not possible to know what is correct, so restrict attention to ensuring that the recorded values are plausible. Nor can you expect to spot all errors, although hopefully you will find the major ones. However, if there is no reason to suspect that seemingly strange, but possible, values are wrong, you should not modify them. Such checking should be carried out as rapidly as possible; the longer you wait, the less chance you will have of being able to obtain a correction. Further data checking and screening will be discussed in Section 4.2.

The database system will then finally produce one or more computer *data files* containing all of the information collected. In simple cases, these will be of rectangular form, with each line referring to one observational unit and each column to one variable, the standard form that most statistical software requires.

Once the data are entered, store the original forms in a safe place. Also make copies of all computer data files and store them in a separate building, in case of fire or theft. It is wise to do this sequentially at all stages of data entry as well.

### 4.1.2 Computer treatment

Computer manipulation of data has both advantages and disadvantages. Among the advantages are that you can:

- quickly handle vast amounts of data;
- obtain results to high precision;
- try many and varied statistical techniques, some very complex;
- easily present data graphically;
- rapidly repeat analyses after making small changes or corrections;
- calculate new variables from those available.

On the other hand, some of the disadvantages are as follows:

- software may contain errors or, more often, poor or inadequately documented statistical techniques;
- with a wide variety of statistical methods available, you may choose the wrong ones;
- you may use the software without any understanding of what it is actually doing in any given analysis, the black box approach;

- if the data or the questions are erroneous, the software will still produce a seemingly plausible answer – garbage in, garbage out.

Massive computer analysis will never be a substitute for clear thought.

Spreadsheets may be suitable for data management, but are not designed for statistical analysis, often even yielding incorrect results.

The criteria for choosing among statistical software include whether they have the following features:

- clear documentation;
- clear, self-explanatory output;
- flexibility for reading data from files;
- data management, such as editing;
- a reasonable maximum amount of data accepted;
- accuracy, precision, and speed;
- a wide choice of appropriate statistical methods;
- treatment of missing values;
- a variety of high-resolution graphics;
- good reputation and reasonable cost;
- good error handling;
- user friendliness and interactivity

(Altman, 1991, p. 110). The relative importance of these factors will depend on the variety of uses to which you will put the software and on the sophistication of the users. Generally, it is preferable to use the same software for all analyses, but, for certain problems, such as clustering, you may require specialized software.

### 4.1.3 Data editing

The goal of data editing is to detect and, where possible, to correct errors in recorded data resulting from data collection and entry. This should involve not only cleaning the data, but also obtaining information on how the errors arose in the first place in order to improve the process during future studies.

In a sample survey, editing may account for 20–40% of the total costs, including computer hardware and software, salaries, and field expenses. Other costs upon which you will find it more difficult to place a monetary value include:

- ill will due to the additional burden on respondents;
- lack of confidence in data quality;
- loss of timeliness of the results due to delays.

Excessive editing can be often counter-productive; you cannot guarantee high quality simply by increasing the number of checks. This may hide problem areas instead of revealing them.

Data errors that you can most easily correct by editing are those that would be recognizable to a user of the individual data records but not having any supplementary knowledge about each given unit. The suspicious items are more difficult. You can judge the latter in relation to their influence on the estimates (Section 4.4.1) to be calculated. Begin with the most extreme suspicious values and stop verifying when further corrections have little effect on the estimates.

The focus of recontacting respondents should be to acquire knowledge of the respondents' problems and causes of errors rather than just to determine that a suspicious value is wrong and to find a more accurate value. This will lead to improvements in your future study designs.

## 4.2 Descriptive statistics

### *4.2.1 Univariate statistics*

The first step in any data analysis should be to produce the simple descriptive statistics for each variable in the database. For categorical data, provide the frequencies and *percentages* in each category. You may supplement these by *histograms* that display this information graphically. If count data involve only a few small values, you can use the same methods for them.

For large counts and for measurements, calculate the *mean*, *maximum*, *minimum*, and *standard deviation*. It may also be useful to categorize the values and to produce percentages and histograms, as above.

Carefully scrutinize all of your descriptive statistics for anomalies. This is the beginning of the second stage of data checking. If you have used a proper entry grid in a database system, no impossible values should be present. Then, any anomalies might consist of unexpectedly large or small percentages in certain categories, unreasonable maxima or minima, and so on, called *outliers*. Any problems that you detect must be traced back to the original forms, or further if necessary, in order to make the appropriate corrections. Again, you should only change values that you are certain to be wrong. If an outlier is correct, it may indicate an anomalous individual, perhaps one who is not actually in the eligible population or one who is of special scientific interest.

### 4.2.2 Cross-classifications

Once you are satisfied that the univariate statistics are reasonable, you can consider the relationships between pairs of variables. For even a small number of variables, this will yield an enormous number of combinations; you will usually not be able to study all of them in detail. Then, select the most crucial and informative ones, especially those related to the main response variable(s).

For categorical data, two-way frequency tables, or *contingency tables*, are generally most useful. Most statistical software will produce percentages for both rows and columns. When numerical quantities are involved, whether counts or measurements, *scatter-plots* are often most useful. Occasionally, non-parametric methods may be helpful in drawing smoothed lines through the data.

Inspection of these results will provide logical checks, revealing impossible *combinations* of values, something that you could not detect by the previous methods. Thus, for example, if you consider age and year in school, three-year-olds in grade five would be suspect. In the same way, the number of previous pregnancies should be undefined for all men. Such inspection may also reveal pairs of values that are individually plausible but impossible in combination.

If you have recorded a series of values, say responses over time, on each individual, plot them, as profiles, to check that they vary in an acceptable way. For example, height of children should gradually increase, with no decreases.

In experimental trials, you should make checks on the degree of equality among randomized groups, especially if they are small. However, testing (Section 4.4) if departures from equivalence are due to chance is irrelevant, unless you suspect errors in the randomization process itself, because you know that allocation was random, and random departures will be present.

Study of these simple descriptive statistics also enables you to gain a first familiarity with the structure of the data. They will provide a basis for much of the presentation of the final report; however, you will interpret them in the light of the more sophisticated analyses that you have performed but that many readers may not be able to understand.

## 4.3 Role of statistical models

The objectives of a study should indicate a few main analyses of primary interest. This will be particularly so in experimental trials. Nevertheless, *exploratory* inspection of the data will generally provide further important information that you should not neglect. However, this is hypothesis generating, not *confirmatory* analysis; the latter will require new data.

Any data collected contain a mass of information. The problem is for you to extract that part of it that is relevant to the questions to be answered by your study, in the simplest and most understandable way possible. This essentially involves checking for pertinent *patterns* and anomalies in the data. This is a basic role of statistical models: to simplify reality in a reasonable and useful way, a way that you can empirically check with the data. No model is ever 'true', but some models are more useful than others for given data and questions.

Models can serve many roles. They can provide:

- a parsimonious description or summary of results, highlighting the important features;
- a basis for prediction of future observations;
- biological or social insight into the processes under study;
- a test of a prior theoretical relationship;
- comparisons of results from different studies;
- measures of precision of quantities of interest.

You can think of models as smoothing the irregularities in the data in a way that makes patterns clearer. The danger, of course, is that a pattern that you isolate in this way is a random artefact of the given data set, corresponding to nothing reproducible if someone were to do a second such study, and hence to nothing in the population under study. The role of model selection, through empirical checking with the data, and of measures of precision is to reduce this risk and to quantify it; you can never eliminate it entirely.

The basic steps in exploratory model building will usually include:

1. studying the pertinent descriptive statistics, as described above, in order to become familiar with the data;
2. developing a reasonable model from the results of step 1 and from previous knowledge;
3. fitting the model to the data;
4. checking the goodness of fit of the model;
5. going back to step 2, if necessary;
6. using the model to draw appropriate conclusions.

The purpose of modelling is not to get the best fit to the data, but to construct a model that is not only supported by the data but also consistent with previous knowledge (unless that is being placed in question), including earlier empirical research, and that also has a good chance of describing future observations reasonably well.

Before looking at how you can actually fit models to empirical data, let us consider some principles of model construction. In constructing a model, the response and the explanatory variables play very different roles. Let us consider them in turn in the next two subsections.

### 4.3.1 Choice of probability distribution

The main response variable that you will study should be that specified in the protocol. In most cases it is directly observable, but in some experimental trials it may be constructed – for example, the difference between the response at baseline, before the intervention began, and the final response after a certain length of treatment.

In statistical models, we consider the response variable to arise *at random* in a certain sense: we cannot predict in advance exactly what response each respondent will give so that random fluctuations are not reproducible. This variability arises primarily from differences among human beings, in contrast to studies in physics or chemistry where measurement error is predominant.

You can then represent the frequencies of the different possible responses by a histogram. This is so even if you are making quantitative measurements, because you can only record them to some finite precision; hence, the observed values are actually categorical.

Unless you have made a very large number of observations, histograms will generally be rather irregular, with a bumpy appearance. If the response consists of unordered categories, such as a list of career choices or of types of illness, the shape of the histogram has little meaning: you can arbitrarily modify it by changing the order of the categories. Then, you have to construct models directly in terms of the probabilities of respondents falling into the different categories. The most common case is a binary response, but a number of models are also available for nominal and ordinal response variables.

If the response is a count or a measurement, you can go considerably further. The shape of the histogram now has a meaning. A *probability distribution* is a mathematical function that smoothes the histogram in an informative way, while retaining, and highlighting, the basic shape. For example, everyone is familiar with the smooth bell-shaped form of the *Gaussian* or *normal distribution*.

An added advantage of such smoothing is that different distributions correspond to different ways in which the data might have been generated. Thus, the normal distribution arises when a large number of unknown small effects add together to generate the response, as for example with genetically inherited traits and multiple environmental influences.

Most probability distributions have one or two unknown and unobservable *parameters* (not to be confused with the observable parameters of the scientist, our explanatory variables). For example, the normal distribution has the mean and the variance. These parameters allow flexibility so that you can adjust the distribution to fit as closely as possible to the empirical histogram. Most distributions have a parameter that indicates the size of the responses, generally the mean. Some have a second parameter related to the shape of the histogram. For the normal distribution, which is symmetric, this is the variance. Most distributions, however, are not symmetric and will have a different second parameter, if they have one at all.

In modern statistics, there is rarely any need for you to transform the response variable to normality. (The major exception would be if you have made quantitative measurements on an inappropriate scale.) Transformations generally make the results extremely difficult to understand: how do you interpret the average of the square root of your responses? A multitude of distributions are available for non-normal data, as is the software to perform analyses with them. Only use transformations of response variables for scientifically valid reasons.

## Common probability distributions

**Binomial distribution** For binary responses, we require a distribution that describes the only two possible events. Generally, the binomial distribution, with only one parameter, the probability, say $\pi$, of the first of the two events, is used.

**Poisson distribution** For counts, we require a distribution to describe positive integers. Here, the Poisson distribution, with one parameter, the mean number of events, say $\mu$, is commonly used.

**Duration distributions** A large number of asymmetric or *skewed* distributions is available to describe durations until some specified event, such as survival (before death). Often a *survival curve* is fitted, based on the *Kaplan–Meier estimates*. In many cases, it is more convenient to study directly the *intensity* or rate of occurrence of the event of interest instead of the time until the event occurs. The most famous example of this is the *Cox proportional hazards model*, widely used in medical studies of survival.

**Normal distribution** This distribution is well known although it is rarely encountered in practice, except as a convenient approximation.

The distributions mentioned are only the commonest ones. For example, frequencies of binary events and counts may show a large amount of variability, called *overdispersion*, that must be taken into account with special distributions. This is almost bound to occur, for example, if clustering is used in a design, as in multi-stage surveys and multi-centre trials.

In contrast to models based on the normal distribution, for all other common distributions the variance cannot remain constant when the mean changes. For example, as we saw when calculating the sample size for counts in Section 2.6.2, for the Poisson distribution, the mean is equal to the variance.

### 4.3.2 Regression models

The probability distribution describes the random variability in the response variable. However, in a study, you will usually be primarily concerned with *systematic* changes in response under certain conditions, the explanatory variables. You can translate this into a statistical model by looking at how the probability distribution of the response, or more exactly the parameters in it, change under these conditions.

As usual, you generally must make simplifying assumptions. Thus, in many circumstances, you may reasonably assume that only the mean of the distribution changes with the conditions of interest. You can take the basic shape, for example as indicated by the variance, to remain constant under all conditions.

A second simplifying assumption is more peculiar to the statistician, not being directly relevant to the scientific endeavour. The way in which the mean varies with the conditions described by the explanatory variables is taken to be linear in the unknown parameters. In contrast, scientists are interested in the (non-)linearity of responses with respect to the explanatory variables (their parameters). The statisticians' linearity is an old historical assumption that was necessary to facilitate computation, and that is no longer required with modern computing power. Unfortunately, most software packages do not meet such modern criteria.

When you combine these conditions, you will obtain a standard (multiple) linear *regression* model, whereby some function of the mean changes with the conditions:

$$g(\mu_i) = \beta_0 + \beta_1 x_{i1} + \beta_2 x_{i2} + \cdots \tag{4.1}$$

where $\mu_i$ is the mean for the $i$th subject, $x_{ij}$ is the observation of the $j$th explanatory variable for that subject, and $\beta_j$ is the corresponding unknown parameter, the *regression coefficient*, to be estimated. This model function, that combines some probability distribution with a linear regression, has come to be known as a *generalized linear model.*

Notice that the function, $g(\cdot)$, is a transformation of the mean, not of the observations, so that it does not produce the difficulties of interpretation discussed above. Your choice for this function of the mean generally will depend on the type of response that you have observed. Common possibilities include:

- *binomial distribution* — log odds or logit: $g(\mu) = \log[\mu/(n-\mu)]$ where $\mu = n\pi$;
- *Poisson distribution* — logarithm: $g(\mu) = \log(\mu)$;
- *duration distributions* — logarithm: $g(\mu) = \log(\mu)$;
- *normal distribution* — identity: $g(\mu) = \mu$.

The logarithm is a particularly important transformation because it allows you to compare means as ratios instead of as differences; effects are multiplicative instead of additive. You can then study relative, instead of absolute, differences. (Note that a logarithmic transformation of the observed responses does not have this simple interpretation.) In linear models, the logarithm also ensures that the mean cannot be negative, often an important requirement. However, a number of other functions are also possible and are fairly widely used.

For a binary response, the logit transformation yields *logistic regression*; this is probably the most widely used regression model of all. It is noteworthy that this is the only regression model that can provide the correct estimates of regression coefficients in a case–control study.

The Poisson distribution with a logarithmic transformation is the basis of *log linear models* for categorical data.

Duration data, such as survival times, present special difficulties that are now widely known. In particular, many observations may be *censored*, that is, incomplete in that the event did not occur before observation had to stop so that the duration is only known to be at least a certain length. These are not missing values or drop-outs and provide essential information about the longest durations. All standard software packages handle such data.

More complex constructions are required for nominal and ordinal response variables because the definition of a mean is not so obvious. These are well documented in many books. One simple way to handle ordinal variables is to assign a known scale to the categories. Care should be taken with this approach, however, because it will be misleading if the scale is poorly chosen.

The $x_{ij}$s in the model need not be simply the observed explanatory variables. In contrast to response variables, here transformations can often be useful. Furthermore, you can handle interactions among variables by including their multiplicative products in the regression. These allow one condition to influence the response in different ways depending on other conditions.

The regression models commonly available in statistical software are linear in the parameters. Skilful use of the transformation function of the mean, $g(\mu_i)$, can yield a limited selection of non-linear relationships. However, certain specialized software is available for more complex non-linear models. Use these when they make scientific sense. One common example is an unknown transformation of a variable, such as $x_{ij}^{\alpha}$ where $\alpha$ is an unknown parameter.

It is often forgotten, when analysing the usual multiple regression models, that the fact that no relationship is detected in a such a linear model does not exclude the possibility that a non-linear relationship is present.

## Complex sample designs

Take great care when you have used a design involving clustering, as with multi-stage surveys and multi-centre trials. You can only apply standard regression models as a rather poor approximation. The model selection procedures to be described below will generally include too many explanatory variables in the final model and the precision of the parameters will be overestimated. Instead, you will require special software that is not widely available, using random effects models, to take into account the dependence among responses within each cluster. This will allow you to make more reasonable model selection and precision estimates.

You can handle more simply the case of stratification with sampling fractions differing from the population proportions. Reweight the observations in the strata to bring them back to the population values. Good software will handle this fairly automatically, once you have calculated the proper weights.

When you have found the appropriate model for dependence within clusters and/or for stratification, handling and interpretation of the regression aspects is not fundamentally different from the simpler cases.

## Interpretation of explanatory variables

**Binary variables** You can usually most usefully code a binary variable, $x_{ij}$ in Equation (4.1), using zeros and ones. However, pay attention as to which of the two possible values has the value one. The category with the value zero is often called the *baseline*, because you will make comparisons to it. Then, you can interpret the corresponding regression coefficient, $\beta_j$, as a *contrast* between the categories, that is, the difference in the transformed mean for the category coded as one, as compared to that for the baseline coded zero.

**Several categories** If the values of a variable correspond to several categories, such as marital status or religion, your interpretation in terms of contrasts will be similar to that for binary variables, but somewhat more complex. Coding may be as alphabetic names or as numbers. In the latter case, remember, and indicate to the software, that these numbers are simply codes for categories and not measured magnitudes. Otherwise, the software will generate completely erroneous results. Generally, the software will produce an automatic recoding to a series of *dummy variables* indicating in which category each individual belongs.

Again, with most software, you must choose one category as the baseline to which you will compare the responses in all other categories. The choice will not alter the final interpretation. You should make it for convenience in interpreting the contrasts among categories. Here, there will be a set of regression coefficients, one less than the number of categories, the missing one being the baseline category. You can interpret each parameter in the same way as for binary variables, as the difference in the transformed mean for that category, as compared to that for the baseline.

You can directly compare the sizes of the coefficients, with larger values indicating greater mean differences from the baseline category. If some values are close to zero, the corresponding categories may possibly be collapsed into the baseline category. If the values for two or more categories are similar, they may also be combined together. These steps will simplify the model by reducing the number of parameters.

**Quantitative variables** When an explanatory variable, $x_{ij}$, is a count or measurement, the corresponding regression coefficient will be the *slope* of a straight line describing how the transformation of the mean changes per unit change of that variable. For this reason, the size of the coefficient will depend on the unit of measurement. For example, if the variable is measured in centimetres, the coefficient will be smaller than if it is measured in metres. Thus, in contrast to the previous cases, here the sizes of the coefficients are not directly comparable.

**Interactions** In simple cases, you can represent an interaction in a regression model by the product of the two (or more) explanatory variables concerned. This indicates that the response depends on any one of these variables in a different way depending on the value(s) of the other(s). Thus, for example, when treatment and sex interact, this might mean that treatment has a larger effect for one sex than for the other. When a variable with two or more categories interacts with a quantitative variable, this indicates that the slope for the latter variable is varying among the categories of the former variable.

In all cases, when several variables are present in a regression function, each coefficient gives the dependence of the (transformed) mean response on the corresponding variable, *conditional on the values of all other variables in the model.* Thus, your interpretation of each given coefficient will change every time you add another variable to the regression model, or remove one.

Strictly avoid composite indices constructed from a set of explanatory variables. Although they have often been used in practice, they have several drawbacks:

- Considerable information will be sacrificed, especially if an inappropriate index is constructed; in bad cases, the result may be worse than using only one of the individual explanatory variables.
- Objectivity is lost because someone else might choose another combination to create the index.

Use of an index in a regression model is equivalent to assuming that the values of all the coefficients of the variables contained in the index are known (up to a constant of proportionality). It is preferable to estimate them in the model and to eliminate those variables that are unnecessary.

### 4.3.3 Multivariate models

You will require a *multivariate distribution* when you wish to study the inter-relationships among two or more responses.

This may occur in at least two situations: you may have observed:

1. two or more distinct responses, and it is important to understand the relationships among them;
2. the same response several times on each subject, usually at successive points in time in a longitudinal study, known as *repeated measures*.

The most commonly (mis)used parameter to study such association among responses is the *correlation coefficient*. This resembles a regression coefficient for dependence of a response on explanatory variables, but has a very different interpretation. Except as an approximation, a correlation is generally restricted to interdependence among responses having normal distributions, something that is not true for a regression coefficient.

Many techniques are available for studying multivariate relationships, in the first context, such as discriminant analysis, principal components analysis, and factor analysis. Most do not allow the introduction of explanatory variables and so are of limited use in the applications being considered here. They are most often used as descriptive tools for exploratory analysis.

Much more sophisticated multivariate models are becoming available for repeated measurements, although the software is only slowly appearing. Generally, here you must consider two types of dependence among responses on a given individual:

1. *serial dependence*, whereby those observations closer together in time are more similar;

2. constant *intra-subject dependence*, whereby all observations on one subject are more similar than randomly chosen observations from different subjects.

For the former, never forget that human beings have memory and the ability to learn. In your models, you must properly condition on the previous history of each individual. For example, subjects who first meet a task while tired may continue to do it badly when rested, whereas those who first encounter it when fresh may go on doing it well when tired.

The second form of dependence arises because the same subject tends to respond consistently in the same way, always with high or low scores on tests, high or low blood pressure, and so on. In other words, subjects will be heterogeneous on the many unobserved confounding explanatory variables.

A special case of such models is that for clustered designs. Here, there are several levels of analysis, from the topmost cluster to the final observation unit. The special software, mentioned above, takes into account the multivariate dependence among what are in fact 'repeated' responses within clusters.

## 4.4 Model selection

Until now, we have been considering possible statistical models without actually *fitting* them to the data. Model selection involves two phases: choice of possible models to be considered, before the data are collected, and choice among these models in the light of the data, possibly even finding or developing new ones if necessary. The latter choice is our next task.

The actual procedure depends on the context. In decision-making situations, you must have a clearly defined set of alternatives available from which to choose. On the other hand, science fundamentally involves attempting to discover the unexpected. Only subsequently is replication used to confirm what was found.

When you do exploratory analysis, during preliminary research, you still only know a little about the phenomenon under study so that you will often have a wide variety of models under consideration. Subsequently, you may construct clear hypotheses, based on the information you have obtained, for which you will then need to design a study involving confirmatory analysis. Model selection procedures are logically different in the two cases.

The first phase, that we have been discussing up until now, is largely determined by the way in which you designed the study. You cannot consider a model for which you have not collected appropriate data. For example, you cannot use a model containing explanatory variables that you have not observed.

If you are legitimately to apply any tests, you must clearly specify the corresponding hypotheses, as models, in the protocol before the data are collected. Thus, if you conduct some experiment with treatment and control, you can test a model with no difference between the two against a model with some scientifically large difference, usually in a specific direction.

Most often, you cannot specify such clear hypothetical models prior to a study, except in advanced stages of research. This is especially true for sample surveys where you collect a large number of explanatory variables. Then, you can apply a model selection procedure empirically using some criterion based on the observed data, moving from one model to another in a way that cannot be foreseen before collecting the data.

### 4.4.1 Parameter estimation

Once you have some model function under consideration, one of the first things that you generally want to do is to obtain, from the observed data, *estimates* of the unknown parameters in that function. Thus, a first simple step of model selection is to choose among parameter values for a given model function.

One main role of statistical models is to provide a means of calculating the probability of observing various possible responses in a situation of random variability, as described above (Section 4.3.1). We can invert this reasoning to provide a criterion for the relative plausibility of various models, that is, a means of selecting among them. We can say that one model is more plausible or likely than another, in the light of the given observed data, if it makes those data more probable. This procedure yields a function of all possible models, giving the probability of the observed data for each, called the *likelihood function*. It is the basis of most of modern statistics. Most often, the logarithm of the likelihood is used for ease of interpretation. Thus, a statistical model may be more preferable if it has a relatively higher likelihood or log likelihood, but this will also depend on the complexity of the model and on its relevance to the question at hand.

For a given model function, there is usually one set of

parameter values that makes the likelihood function largest. This set is known as the *maximum likelihood estimate* (m.l.e.) of those parameters. In most cases, it is the intuitively obvious estimate:

- For a mean parameter, the m.l.e. will almost invariably be the sample mean.
- For the probability of a characteristic, what statisticians call an event, it is the proportion of those in the sample with that characteristic.
- For classical linear regression, the least squares estimates are the m.l.e. when a normal distribution is assumed.

All software should provide such estimates, but you must take care, because other estimates may also be available and may even be the default ones given.

### 4.4.2 Selection criteria

In very simple designs, such as an experimental trial, with one active treatment and a control, where a response is observed at one end-point, selection among model functions may not be necessary. You only wish to determine if the observed mean difference in response between treatments, for a given known probability distribution, could readily have arisen by chance if there was no difference between them.

Often, however, even in this simple situation, you will not be prepared to assume that treatment effects remain constant over all individuals. Then, to control for these differences among individuals, you must introduce into the model additional variables, in this context often called *covariates*, obtained from the concomitant observations foreseen in the design. One very particular example of this occurs in cross-over designs where the use of the previous treatment as a covariate may allow for the effect of carryover.

In more complex situations, you will not know the appropriate probability distribution and you will have several explanatory variables, even a large number in many sample surveys. The use of different distributions and the inclusion of different combinations of the variables correspond to different model functions. Thus, you will require means of selecting among them.

The common way of performing model selection for regression, by testing if the coefficients are significantly different from zero, is simply wrong: hypotheses for tests must all be constructed before looking at the data. This is impossible in any step-by-step model selection. The standard probability levels used are incorrect, and the correct ones impossible to calculate. In addition, you can obtain contradictory results, depending on how many parameters you test at each step, even if done in the same order.

The use of the likelihood function in estimation is straightforward and well known. However, the problem with the likelihood function as it stands, for selection among model functions, is that it does not take into account the complexity of the functions being compared, that is, how many parameters each one has. The more complex is a model function, the more chance it has of making the data probable. But you will generally prefer simple models:

- they are more easily understandable;
- they can be more readily communicated to colleagues, es-

pecially to non-specialists;
- they have more chance of being reproduced if the study is repeated;
- they often provide superior predictions.

Thus, in model selection, you will need to balance the two: high likelihood and simplicity. You can do this by penalizing the log likelihood by subtracting some function of the number of parameters involved.

In this way, proper model selection criteria have been elaborated in recent years. The best known and most widely used is the *Akaike information criterion* or *AIC*. You can obtain it simply by subtracting the number of estimated parameters from the log likelihood, thus slowing down the rate at which the likelihood increases as parameters are added. Take care, however, because most software packages give the negative of this value (often multiplied by two) so that small values indicate better models. Notice that the actual value of the AIC for a given model has no meaning, only its size in relation to that for another model fitted to the same data. Model selection criteria are relative, only allowing comparisons among models.

### 4.4.3 Selection strategies

Especially in a sample survey, you will often collect information on a considerable number of explanatory variables. You must elaborate some strategy, generally specified in the protocol, in order to find your way through the maze of possible models.

Many software packages provide automatic variable selection procedures, such as forward and backward stepwise techniques. They can rapidly fit an enormous number of statistical models and provide masses of output. Avoid this approach for at least three reasons:

1. It does not take into account the relative importance of the variables, as known to workers in the field.
2. It generally uses classical tests and not proper model selection criteria, such as the AIC.
3. The more models are tried, the more will describe the data well just by chance.

Instead, your strategy should be to use all available information about what could be reasonable models so as to minimize the number considered. You should thus specify the general order in which the variables will be entered into the model before looking at the data.

In most studies, you can classify the explanatory variables into a number of groups depending on how closely they are associated with the unit of observation. Thus, for example, if children are being studied, you might have:

1. variables directly describing them, such as sex, age, household activities, health, and so on;
2. variables describing the household, including the parents, the house, and so on;
3. those for the community, such as size of village, distance to a shop, and so on;
4. those for the whole region of the country, such as climate and geography.

A reasonable strategy will often be first to study those variables most closely related to the observation unit to see how much of the variability among responses can be explained by them, before adding in variables from the next set, and so on. Remember that all parameters in a model change interpretation as each new variable is added.

Within a set of variables, you may use various strategies. You may add each one according to some order of importance, if this is possible. Or you may enter the whole set, and then withdraw the unnecessary ones, one at a time. The former is often preferable because you are then starting with the simplest and most important explanation of the observations.

In observational studies, take care with pairs of variables that are providing very similar or identical information. Such variables are said to be *collinear*. If you include either one of such a pair of variables in a model, you may find it necessary, but if you include both, neither may be indicated as necessary. Then, make some choice as to which is more pertinent, and only keep one.

Also, especially in observational studies, take particular care with extreme values with respect to any given model, called outliers. Each individual observation point may provide you with a clue to explaining the observed associations. Values lying far from the model, that is, having large *residuals*, can indicate ways to improve the model, but they may also simply be rare values.

## Missing values

When you begin the task of fitting and comparing models with computer software, it is essential to indicate how missing values are recorded for each variable. If you have coded the missing value as 99, whereas plausible values range from zero to 20, the software will use the missing value code in the calculations unless you have told it that that value is special.

Most software will simply eliminate missing values from the analysis. The implicit assumption being made is that the missing individuals are representative, that is, that the values are missing at random. However, when studying human beings, we know that this is almost never the case. If you have available some information about the reasons for missingness, you may incorporate this into the model, although this is usually a difficult process.

Take special care with complex models containing a number of explanatory variables. As you add variables in the selection process, each may have missing values on different individuals so that the number of observations actually used can diminish drastically. This generally also means that the bias is rapidly increasing and the results can be completely misleading.

In the present state of statistics, you can do very little about non-random missing values without making major assumptions. One possibility is to construct binary response models to see how missingness depends on the other available variables, but this will be of little help in correcting for missing values in a model for the response of interest. Thus, be very careful about the accumulation of missing values in complex models containing many variables.

## 4.5 Estimating precision

As we have seen, model building involves estimating unknown parameters. However, you will also require some idea of how precisely these parameters can be calculated from the information in the data.

The way in which you can estimate precision depends on what type of model selection you performed. If you had only one known model function before starting, no major problems arise and you can use classical forms of indicating precision, such as significance tests and confidence or credibility intervals. However, if you first apply some step-by-step selection procedure among model functions, the uncertainty in this process will not be taken into account in such classical procedures and the stated probability levels will be incorrect.

Recall also that precision will be overestimated if you use standard software for designs that involve clustering. Because of the dependence among responses, the sample size of observational units is not equivalent to one without clustering. The data will contain considerably less information than a simple random sample of equal size and, thus, you should not treat them as such.

Finally, if you did not fix the sample size in advance, but determined it in some sequential trial using interim results at each stage, the true probability levels of classical significance tests and confidence intervals will require non-standard methods and be complex to calculate.

### 4.5.1 Significance tests

In simple cases, especially in experimental trials, you will have some specific hypothesis in mind that you want to test by the data obtained from the study. In the protocol, you would generally have stated this, in relation to the primary end-point, as a *null hypothesis* of no effect, for example of some treatment, on the observed response.

Intuitively, you are asking if the observed differences in response under two or more conditions (defined by the explanatory variables) could have arisen at random if those conditions had no effect. If the probability of such a thing happening is low, often taken to be less than 5%, either the data represent an extremely rare occurrence or the null hypothesis is wrong. Such a probability is known as a *P-value*. Thus, statistical significance is a measure of the reliability of an observed relationship.

No test is possible if no model is formulated for the data. Only a model can provide the probabilities that tell you if the observations are rare or not.

Common tests include:

- the *Student t test* for an individual parameter;
- the *F test* for a set of parameters;
- the *Chi-squared* ($\chi^2$) *test* for one or more parameters and for goodness of fit;
- *Fisher's exact test* for contingency tables.

For most tests, like the first three just mentioned, you must make the assumption that the model is based on or can be approximated by a normal distribution. (In other cases, the statistic is assumed to have the appropriate asymptotic distribution; that is the sample is assumed to be very large.)

Such procedures are generally acceptable *when you have clearly stated the null hypothesis in the protocol.* Note that, in reporting the results, you should state the more informative exact P-value, and not simply an indication as to whether it is less than some fixed level, such as 5%.

In terms of a model, what you have done is to calculate how rare it would be to observe the estimated difference in responses among treatments if the actual difference were zero. In other words, if the parameter of interest, describing treatment difference, is zero, what is the probability of estimating it to be as large as that calculated? Thus, such a significance test is a crude measure of precision of the parameter estimate.

In many scientific journals, significance tests are actively discouraged, if not forbidden, in favour of some form of confidence or credibility interval for the estimated difference parameter.

### 4.5.2 Confidence and credibility intervals

Many studies are more exploratory and do not have one such simple hypothesis to be tested. As a next stage of complexity, suppose that some one specific model function is of interest and that your goal in the study is to estimate the unknown parameters in it. The best estimates will be the maximum likelihood estimates, following the criterion given above. However, there is little chance that these are the actual values in the population. Thus, you require some measure of precision.

As we have just seen, one simple procedure would be to test if each parameter were zero, obtaining a P-value for each. This is only acceptable if you have clearly stated both the model, and the parameters to be tested in it, in the protocol.

A more widely used and preferable procedure is to calculate confidence or credibility intervals for each parameter (or some set of them). The common procedures include:

- giving the parameter estimate, plus or minus a fixed number of standard errors;
- calculating an interval based on the log likelihood, perhaps penalized by a *prior distribution* of the parameter, obtained from previous knowledge or beliefs (known as the Bayesian approach).

For the classical normal linear models, the two will generally give identical results (unless prior belief is strongly contradicted by the new data). In other cases, the likelihood-based method is preferable, because the intervals should not be symmetric around the m.l.e. Standard errors can then be very misleading, even including impossible values in the interval. However, such likelihood-based intervals are more difficult to calculate and are rarely automatically given by standard software packages.

Such intervals have a probability level assigned to them. In the same way as for significance tests, such levels only have meaning if you have specified the exact model function for which they are calculated in the protocol. If you chose the model function by some selection criterion, using the data, the uncertainty in that selection process will not be allowed for and the intervals will be too narrow for the stated probability level.

### 4.5.3 Likelihood intervals

In a context where you have used model selection to arrive at a final choice, the only way to provide objective intervals of precision for the estimated parameters is directly through the likelihood function. Thus, for example, all parameter values that make the data, say, one-tenth as probable as the maximum likelihood estimates will yield such a likelihood interval.

In fact, such intervals will be identical to likelihood-based confidence or credibility (with an uninformative prior) intervals discussed above. The main difference is that, in the present context, you do not know what probability level to assign to them. You can only judge them in terms of the relative likelihood of the parameter values given the data. (You can think of the interval obtained by using a standard error as a quadratic approximation to such a likelihood interval.)

Thus, confidence or credibility intervals, calculated for the final model obtained after some selection procedure, will provide measures of relative precision of the remaining parameters. However, you can only measure the level of such precision statements in terms of the (relative) likelihood itself and not by their nominal confidence or credibility probabilities.

If you leave unnecessary variables in a model, instead of eliminating them by the model selection procedure, the precision of parameters of interest will generally be less than if you eliminate them. However, if you specified that model in the protocol, you cannot eliminate these variables without destroying the probabilistic basis of the intervals of precision for the parameters of interest.

Thus, tests and confidence or credibility intervals are primarily of use in decision making, as well as in confirmatory scientific work, whereas likelihood intervals will be used in the exploratory phases of scientific inference.

## 4.6 Summary

Transfer all data collected to electronic form, preferably using a database management system. Each of the basic types of observations – counts, measurements, and discrete categories – has its own specific potentials for errors. You should detect, and correct, as many errors as possible at this stage.

You will need to choose appropriate computer software that is reliable, can perform the required analyses, and is reasonably easy to use.

The first step in data analysis will be to produce descriptive statistics such as frequency or percentage tables and histograms for qualitative variables and means, standard deviations, maxima, and minima for counts and measurements. You can then cross-classify the most important variables in frequency tables or with scatter-plots. Check all of these for suspect values and correct them, where possible.

You can use statistical models to search for informative patterns in the data. In such models, you assume that the response variable follows a probability distribution. They will have unknown parameters that you must estimate from the data.

Usually you will be interested in how such a distribution changes under the different observed conditions. You can do this by allowing one or more parameters in the distribution to vary with these conditions, the explanatory variables, in a regression model. Binary and qualitative variables provide contrasts with some baseline condition, whereas quantitative variables describe how the response distribution changes per unit change in such a variable. Interactions allow one variable to have a different effect depending on the value of another variable.

You will require multivariate models when you wish to analyse several response variables simultaneously. An important case occurs when you make repeated measurements on each subject.

Model selection includes both choosing possible models for consideration before beginning the study and selecting among these, or possibly adding new ones, in the light of the data. The likelihood function tells you how probable your data are for each model, thus giving you a measure of their plausibility. In this light, for a given model function, the best parameter values are the maximum likelihood estimates.

Stepwise testing of models, for example in multiple regression, is wrong because it does not adequately allow for the uncertainty involved. Use proper model selection criteria, such as the Akaike information criterion. Instead of using the automatic (stepwise) variable selection procedures available in software, you should organize your model selection strategy in a scientifically sensible way. Take special care with missing values, although there is no easy answer.

The classical way of estimating precision is by means of tests. However, these carry the important assumption that the models were specified in the protocol. Report actual P-values instead of just comparing them to some fixed level. However, a preferable method is to use confidence or credibility intervals for the parameter estimates. If you did not state the specific model you are considering in the protocol, the only objectively valid intervals of precision are those based directly on the likelihood function.

# 5

# Reporting the results

## 5.1 Evaluation of the study

### *5.1.1 Decision making and scientific inference*

The studies with which we have been primarily concerned aim to produce information to further scientific knowledge or to aid in making decisions as to some course of action. Because this information must be as objective as possible, I have emphasized the importance of the accuracy and precision of the results. However, the statistical material you accumulate will never be sufficient basis for making a decision in itself. The role of the statistician is to provide objective information, not to tell the people in policy-making positions what decisions to make.

The decision-making process is always complex:

- Decisions are rarely simple. They depend on information from several types of observations, not just statistical ones, and on the relative costs of the alternatives.
- The final decision is often an act of judgement, weighing political and commercial factors, as well as scientific and technological ones.

Formal decision theories, including statistical ones, are almost universally designed to explain the choices of a rational individual; they are unable to take into account complex collective factors. They are generally of limited use as an aid to decision making within industry or government.

Nevertheless, simply producing objective results directly applicable to such a decision is usually unwise. Generally, it is advisable also to attempt to reach some understanding of the underlying process that you are investigating, rather than only recording, say, the magnitude of some contrast between treatments and its significance level. In addition, the results of a study can often be useful in unexpected ways, helping to resolve questions different from those specified in the protocol. You should present the results in a form allowing for these possibilities. Science is an ever-continuing process, relying on a community of critical minds judging results on the basis of shared information.

### 5.1.2 Criteria for assessment

Once you have collected and analysed the data, you should critically assess your study. It is meant to convince others of the validity of your conclusions; they certainly will closely scrutinize it. This evaluation will include (Elwood, 1988, p. 164) the following:

- Description of the evidence:
  - What is your study design, is it appropriate to your objectives, and how well did you follow it?
  - Does it involve an intervention?
  - Was your sample size adequate to detect the relationships you were looking for?

  - What are your eligible and target populations?
  - How did you choose the subjects involved?
  - What are the rates of missing values and, in an experimental trial, of compliance, and what biases may they introduce into the conclusions?
  - What relevant explanatory variables do you have and how accurately did you measure them?
  - What is your outcome or primary end-point?
  - How can you summarize your main result?
  - How appropriate are your (absolute or relative?) comparisons among groups?
  - Have you provided adequate measures of precision of estimates?
- Internal validity:
  - Is there an effect of observation bias?
  - Are all necessary explanatory variables available or could there be confounding?
  - How large is the effect of random variation?
  - Is there a correct time relation?
  - How strong is the relationship between source of exposure and outcome?
  - Does the outcome vary with the magnitude of the source?
  - Are your results consistent within your study?
- External validity:
  - Can you apply your results to the eligible population?
  - To the source population?
  - To the target population?
- Comparison with other evidence:

– Are your results consistent with those from other studies, particularly similar or more powerful ones?
– Are your results plausible?

In summary, the most important point is that you made all possible efforts to eliminate biases at the design stage. The main ways that you have to convince people that you have done this objectively are:

- to set out your project for the complete study clearly in a protocol before beginning;
- to randomize wherever necessary and possible;
- to use appropriate instruments;
- to take all possible measures to eliminate missing observations;
- to carry out analyses appropriate for the data; and
- to draw conclusions justified by the information available.

Design is unquestionably the most important aspect of the study, because alternative analyses can always be performed once the data are available, but it will be too late to change how the data were collected.

When administrative staff are heavily involved, you may need to take extra care about the quality of the results. Certain bureaucrats tend to be more concerned with the internal coherence of results than with their relationship to empirical reality.

## 5.2 Interpreting the results

Statistical analysis is based on comparisons, the art being to choose the relevant ones for the question at hand. Then, once you have observed a comparative difference, you will have several candidate explanations available. The one explanation that is virtually unique to statistics is that the difference simply arose by chance. The first role of your study is to provide supportive evidence that such an explanation cannot be sustained.

However, this may still generally leave you with a variety of other believable explanations as possibilities. Thus, a well-designed study should also provide evidence to permit all but that of interest to be convincingly dismissed in the interpretation of the results. In other words, it must be possible to present the results of your study in such a way as to convince people that they did not arise by chance or by some unspecified factors not taken into account.

### *5.2.1 General principles*

Chance is at the very foundation of statistical reasoning. This can lead to certain illusions. Thus, relationships are rarely deterministic, so that exceptions never prove the rule in statistics. Non-smokers can have lung cancer; adults without formal education may be learned. Such cases demonstrate that confounding factors are operating that have not been unaccounted for, although they may usually be much less important than the main factors. It is also commonly thought that 'the errors will balance out' in statistics. This will never be true for errors of bias, as when a study is incorrectly designed. Random sampling errors have decreasing importance as the sample size increases; biases do not.

Average values are not necessarily normal or typical values, and will often be impossible: no family has 2.3 children. The variability is essential. A mean only provides an indication of the location of response values around which variability occurs, not of what the response should be. In the same way, a regression line cannot be interpreted as meaning that most or all individuals would be expected to lie on it; they will not. For example, in normal linear regression, the estimated constant variance (mean residual sum of squares) of the model provides the estimate of the variability of the observed individuals around the line.

Random processes generally tend to produce fairly smooth single-peaked distributions of responses. Multi-modal distributions and outliers often indicate a mixture of several processes that you have not properly distinguished, for example because you have missed important explanatory variables. In the same way, a range of 'normal' responses will often be difficult to establish because the range of abnormal responses will generally show considerable overlap with it.

Take special care with testing procedures.

- Check that you only test hypotheses stated in the protocol. Otherwise, use model selection procedures.
- Be wary of exceptionally small or large test statistics. The former, indicating the model fits too well, will raise suspicion of cooking the results, but can arise from design errors. The latter may often be calculation errors unless the sample is too large.
- Rejecting a null hypothesis does not mean that the alternative is true; a P-value is only a measure of doubt about the hypothesis tested, providing no reasonable indication of a better one.

- Statistical significance depends on sample size, not on scientific importance; with a large enough sample, you can detect the most uninteresting minimal difference.

Testing is generally misused: any idea that occurs to the research worker during analysis of the data is tested. However, you must formulate hypotheses or models to be tested in the protocol. If you have, the study is confirmatory, and strong conclusions can be drawn about the relationships you have found. Otherwise, it is only exploratory, and you will require further studies for confirmation of any relationships discovered.

Other miscellaneous points of confusion and error in presenting results include the following:

- Correlation is not a general term to describe relationships but a specific technical statistical term; use association or dependence instead.
- Be careful of the difference between the terms 'parameter' and 'variable', especially when the audience will be a mixture of scientists and statisticians.
- Figures in a computer seem very objective; do not forget the long trail of their human origin.

### 5.2.2 Design dependencies

A first criterion for interpreting your results is whether you have used an intervention or not. Only in the former case can you draw objective empirical conclusions as to causality. In the case of an experimental trial, you must then consider to what extent your subjects are representative of some larger population. This is a question that does not arise for properly conducted sample surveys, as long as the number of missing values is minimal.

In an experimental trial, it is common for the difference in response between two treatments to remain relatively constant even when substantial differences in the mean responses are possible. This often provides one argument for the plausibility of generalization to a larger population. On the other hand, the special attention devoted to a trial may mean that the treatment is quite unsuited to routine use.

A second criterion is generality of the results. This is related to replicability. If the study were repeated, would the same or similar results be obtained? Standard statistical measures of precision are intended to indicate how reliable will be the results with replication under identical conditions. However, generally such replication in identical conditions is impossible in the study of human beings. Thus the question widens to determining in what contexts similar results could be obtained.

Often, you will have included a certain variability of context in the study. This is indicated by stratification in surveys, and blocks and classification factors in experimental trials.

Depending on the goals of the study, the range of contexts of interest and applicability will vary:

- Can you assume that the phenomenon you have studied will be reasonably constant over time?
- In an experimental trial, do the artificial clinical or laboratory conditions inform you about possible results in a more natural setting?
- Will your conclusions be applicable to other subgroups of

human beings than those you have studied, such as other cultures?

When you have included a variety of contexts in the design of a study, it is important to look at interactions between them and the explanatory variables of particular interest, such as treatment in a trial. Do these variables have the same relationship to the response in all contexts? A second possibility is that responses within each particular context show dependencies, as in clustering, so that you may require a random effects model.

It is essential to compare your results with those previously obtained elsewhere. These will generally provide a wider set of contexts than are available in any one study. However, be careful because only studies yielding statistically significant results are generally published, yielding a *publication bias* in favour of relationships among variables.

If you have included the appropriate 'standardized' treatments in an experimental trial, compare these responses with those for previously established results in the field.

One convenient method of comparing all studies on a subject is through an overview or *meta-analysis*. This involves collecting information on all known comparable studies, ideally both published and unpublished, to avoid bias. Then, you will be able to apply statistical modelling techniques similar to those for multi-centre trials or multi-stage surveys. Because of the larger overall sample size, you

- may detect relationships that were not discovered in the individual studies;

- should obtain an improved estimate of effect size, and its precision;
- may resolve disagreements among studies;
- may answer questions not posed at the beginning of the individual trials.

In performing such an analysis, take care not to obscure differences among individual studies.

Never assume that, if some context variable has not been studied, it has no effect on the response.

### *5.2.3 Causality*

#### Experimental trials

Randomization of treatment assignments is supposed to guarantee that you can draw causal conclusions in an experimental trial. This can, nevertheless, be questioned in a number of ways.

In trials where you cannot use blinding of the investigator, your results can be accused of experimenter bias, including bias in the statistical analysis. One precautionary measure is to separate the experimental manipulation from the measurement of the response, blinding the latter as to what occurred in the first step. If no other procedure is possible, you may want to make available filmed records of the interventions.

When human subjects are involved in an intervention, and their blinding is not possible, they may react by:

- trying to please the research workers;
- responding in a socially desirable way;
- regarding the experiment as a personal test of knowledge or character.

This is closely related to compliance with the treatments being applied.

If you do not find a causal relation between treatment and response, this may be because the intervention occurred too late. The process being treated, such as a chronic disease, may have been at too advanced a stage to be reversible.

## Sample surveys

The dependence or association that you find between variables in a model does not provide direct empirical evidence of causality unless your study was an experimental trial. In observational studies, you can only support causality on non-statistical grounds. However, the question of causality will inevitably arise in observational studies, if policy decisions are to be made. Important points to consider include:

- the temporal order of the association;
- the strength of association between source of exposure and response;
- the consistency of association among different circumstances, people, places, and times;
- the specificity of the response to that association;
- the changing strength of association with changes in the strength of the source;
- the scientific plausibility of the explanation.

Consider again the problem of linking cancer to smoking, discussed in Section 1.4.3. A critic might first argue that the cause is not smoking but the tar and nicotine. This, however, would require a demonstration that smoking, as we know it, is possible without them. Smoking and cancer might both be causally linked to some forgotten third variable. You can often rebut such arguments by examining the appropriate subgroups; unfortunately, there is no end to the potential list of such variables.

Where possible, the most powerful strategy, as suggested in Section 1.4.3, is to spell out the postulated causal mechanism in detail before beginning the study. Thus, tobacco smoke contains substances that can cause cancer by contact. This has the following implications:

- The risk of cancer should increase with:
  - the length of time a person smokes;
  - the number of cigarettes a person smokes in a given period of time.
- Specific behavioural patterns should alter the risk:
  - stopping smoking should lower it, or at least stop it from increasing;
  - using filter-tipped cigarettes should lower it;
  - non-smokers in frequent contact with smokers should raise it.
- Cancer should tend to be localized to contact areas:
  - cancer tends to occur in the lungs;
  - pipe and cigar smokers who do not inhale have high rates of lip cancer.
- Finally, smokers should have higher risk of other respiratory diseases.

If you could demonstrate all of these relationships, then they would provide a coherent whole yielding a strong argument even although you have no direct empirical evidence of a causal relationship available.

Paradoxically, different results from a different type of observational study will not refute the original evidence for causality, but the same results from the same form of study will not invariably strengthen the evidence. When summarizing the results of such a study that shows an association consistent with a causal hypothesis, list and discuss all alternative explanations of the results. An observational study can only end with an opinion about causality, not a proof.

## 5.3 Writing the report

### *5.3.1 Presenting the case*

Good statistics involves principled argument that conveys an interesting and credible message. If your report is not interesting, no one will read or talk about it. Thus, it should not be pointlessly formal, concentrating on hypothesis tests, but rather make appealing claims based on intelligent interpretation of the appropriate evidence from empirical observations. Your report will be scientifically interesting if it has the potential to change what scientists, or decision makers, believe about the phenomenon under study (Abelson, 1995).

For this to be possible, your report must be:

- reasonably brief;
- clearly written in an appropriate style, avoiding long and technical words;
- well organized and coherent;
- supported by appropriately informative tables and graphics where necessary;
- adequately referenced;

- free of errors.

Reading your report should be a stimulating and satisfying experience.

Scientific beliefs can be modified by strengthening, weakening, or altering existing ones or creating new ones. Research should be surprising in order to have the potential for such change. For this to be possible, you must show that your research has been conducted without flaws and that your claims are not too incredible to be true. Research issues that are currently considered to be important generally have greater potential to generate change.

The writer(s) of a report must expect that it will come under attack, and be prepared for it. The first line of attack is usually methodological: critics will argue that your design or analysis is flawed. Major criticisms will include:

- lack of or inadequate randomization;
- key missing explanatory variables;
- unrepresentative subjects to allow generalization;

in addition to the usual technical details. Successful rebuttal will depend on providing appropriate details of your research design and your statistical analysis. Thus, your study can only be judged on the information included in the report(s). No reader should be expected to assume that a study was properly designed, executed, and analysed without it being clearly demonstrated in the report. In particular, you must show that you took all possible measures to reduce biases.

Make plans for writing, including individual responsibilities, early, preferably in the protocol. You may prepare a list of possible reports, ranked in order of importance in relation to the aims of your study. Most writing effort will require team work. The team for each report should have a designated chief and should be composed of members with the required expertise. If questions of design and data analysis are involved, you should include a statistician.

### 5.3.2 Basic structure

The basic sections of a technical report are usually the following:

- *title*;
- *material and methods* – the study design and statistical models;
- *results* – analysis and presentation;
- *discussion* – interpretation.

However, it is often preferable to provide a summary of the main results and conclusions at the beginning so that the reader does not need to search for them.

The structure of a scientific paper, essentially similar to that just described, is generally as follows:

- *title*;
- *abstract* or summary of your main results;
- *introduction*, recalling relevant previous research and justifying your objectives in the present work;
- *methods*, describing exactly what you did, that is, the protocol and how you followed it;
- *results*, describing what you discovered;

- *discussion*, interpreting your results and drawing implications;
- *references* to related publications in the field;
- occasionally an *appendix*, containing more technical and detailed supporting material.

All sections, except the discussion that gives the authors' opinions, are supposed to be objective. Each journal has its particular style, with published guidelines; these should be followed.

The first step in the actual writing process will be to produce an outline listing the main points to be covered. It is often easiest to do this in the chronological order in which you conducted the study. You can rearrange the points in a more appropriate didactic order later. From this first outline, connections, gaps, and incongruencies should become apparent. Next, you can compose a first draft.

A major problem in revision is that you will be too close to the material and lack a wider perspective. You can gain some distance by:

- leaving the draft a week or two;
- reading it to someone and seeing where further explanations are necessary;
- having someone else critically read it.

Try to place yourself in the position of the reader. Remember that, *when the reader cannot understand something, it is always your fault.*

Most readers will want to master the main results, and the nature of the argument, along with its limitations. Few will be interested in all the details. They may have to present a summary to their boss or colleagues.

### 5.3.3 *Protocol and design*

Your title should be short, but not cryptic, and to the point, attracting attention to the specific points covered by your study. The abstract is especially important because many readers will go no further, but draw their conclusions from the information in it. Thus, it should contain an accurate summary of the principal findings, with little said of design or of interpretation.

The objectives of the study should be clearly stated.

Present as much detailed information on the way in which your study was conducted as is necessary for readers to judge the validity of your conclusions. This should include:

- your main response variable, or end-point;
- your hypotheses or models, as defined in the protocol;
- how you calculated sample size;
- the types of subjects, including their source, exclusion criteria, and how you selected them;
- how you performed randomization;
- in trials, blinding and controls;
- the types of observations and the measurement techniques that you used;
- numbers of missing values, and any reasons why they might or might not reasonably be assumed to be random;
- in trials, the rate of non-compliance.

For example, it is not sufficient to state that your study used randomization or blinding; you must describe the exact techniques applied. For a survey, clearly state whether it is retrospective, cross-sectional, or prospective.

Provide adequate details of the hypotheses and models specified in the protocol; clearly distinguish these from any that you developed by exploratory analysis in the light of the data.

### *5.3.4 Analyses*

Present statistical analyses in sufficient detail so as to be understandable and so that a reader with access to the raw data could repeat them. Give details of the types of statistical models used and why. Identify complex methods unambiguously, including the software used. It is not sufficient to say that you used analysis of variance or multiple regression. When you applied several techniques, it should be clear which you used where.

Do not analyse repeated measurements on the same individuals as if the observations were independent (Section 4.3.3). In a related context, give details of how you handled any dependence induced by clustering and how you performed weighting if you used variable sampling fractions in a stratified sample.

Give good reasons for excluding any outlying observations.

#### **Estimates and precision**

Give parameter estimates, and measures of their precision, along with a concise description of the model from which they come. Take care that parallel sets of results are not in contradiction, but also that they are not identical (perhaps through copying errors).

In normal models, you can indicate the precision of parameter estimates by placing the standard *error* in parentheses (not $\pm$). In non-normal models, measures of precision will not be symmetric about the estimates so that standard errors are approximate at best and often misleading. Where possible, give likelihood-based intervals.

Do not confuse standard deviations and standard errors. Only use each in its appropriate places. Except for models based on the normal distribution, they are approximations. The standard deviation is a direct estimate of variability in the population, whereas the standard error is a description of the precision of some estimate (that could be a standard deviation).

## Tests

Generally, avoid tests in favour of intervals of precision for parameters. When you do use them, state the type of test statistic. In scientific work, indicating that a test was significant, for example, at the 5% level, without giving the P-value, is unacceptable, although it may be sufficient in some decision-making contexts. However, never present results simply as P-values, without the estimates. The former give absolutely no indication of how strong some relationship is.

A statistically significant result does not mean that it is of any practical importance, nor does it even prove that the relationship is real. Nor does a non-significant result tell you that there is no relationship, only that you have not been able to detect one, that is, that chance may be the reason for the observed relationship.

Do not perform significance tests based on ideas suggested by an examination of the data. For example, testing the difference between the largest and smallest of a set of means is not valid.

If you make enough tests, you will find some to be significant. *Report all the tests made*, not just the significant ones; otherwise, you are committing fraud. For example, in an experimental trial, you may measure the end-point in a number of ways, not all reaching the same significance level for differences between treatments.

### 5.3.5 Results and conclusions

Describe the data adequately, especially the important variables. Never give calculated values to an unreasonable number of decimal places, as compared to the way measurements were made. State the number of individuals used for each calculation. Scatter-plots or histograms may be more useful than descriptive statistics. For profiles of repeated measurements, join the points for individuals over time.

Give the characteristics of non-responders and drop-outs. In experimental trials, thoroughly document and discuss both non-compliance and any side effects from the interventions.

Simple descriptive statistics may serve to illustrate results from more complex analyses that the general reader cannot be expected to understand. (However, clearly document those complex analyses.) Where possible, graphical presentations are usually more understandable than complex equations. This is especially true for survival data where you can plot survival curves. For regression lines, also show the individual observation points.

In experimental trials, describe objectively what happened to subjects on each treatment.

In complex regression models for surveys, compare the dependence relations for each important explanatory variable with the simple dependence of the response on that variable separately. The latter gives an estimate of the direct dependence in the population, whereas the former is conditional on all other variables in the model, that is, 'all other things being equal'. The two may even differ in sign. Clearly explain the meaning of differences between the two.

Your discussion section will provide interpretation of the results of your study, ideally in the context of previous work. The degree to which your conclusions are convincing will depend on a number of factors:

- the *magnitude* of the relationships measured, as compared to their precision;
- the amount of *detail* provided about the observed relationships;
- the evidence for *generality* of application of your results;
- the *interest* that your research can generate, usually by upsetting widely held beliefs;
- the *credibility* of your claims, based on methodological soundness and theoretical coherence

(Abelson, 1995).

The organizers of a study should not lose interest just because no important relationships are found, so-called 'negative' results. *If the sample was sufficiently large* so that a relationship of interest would have been detected if it existed, this, in itself, is an important finding and you should report it.

You can draw conclusions as to possible applications, including limitations of generalizability, and to future research. Justify these from the results obtained! It is always helpful if you frankly discuss the deficiencies of the trial; this lends credibility to your conclusions.

## 5.4 Publication and dissemination

Subject any type of report to rigorous internal review before making it public. You should first:

- check the spelling, punctuation, and grammar;
- verify that tables, graphs, and equations are correct;
- independently have key statistical analyses redone, if feasible;
- double-check information from other publications.

Then, a wider review within the organization should reveal areas of confusion and missing information.

In large studies, you may create a special committee to supervise this process. If possible, distribute the final draft to all members of the investigative group. In certain circumstances, you may also wish to seek external comments.

Once you have completed your study and made your results public, if possible, release your data for research purposes such as secondary analyses. Often, they can be placed in an independent repository with easy access, for example through the internet. However, take great care to conserve the privacy of the individuals involved. In most cases, you can release the questionnaires, data forms, and design details even much earlier.

Publication of a study may lead to requests for supplementary analyses. Pursue any that may shed additional light on your results. If the data are publicly available, you can simply refer some requests back to the requesters for independent analysis.

Publication of results can lead to political reactions, especially if your conclusions are not in accord with the accepted view. Be prepared to face:

- burying unfavourable results in a mass of detail;
- changing definitions;
- discrediting the authors of the report.

In extreme cases, publication may not be allowed in the first place.

### 5.4.1 Technical reports

Many studies are conducted in order to make policy decisions, whether in government or in private industry. The result is often an internal technical report that is not widely distributed. However, the drive to produce valid and objective results may be much stronger than in scientific publishing because concrete actions will be taken. If the results of the study prove to be wrong, the organization involved, and hence also the persons producing the report, will often suffer the consequences, a rather strong incentive.

In certain contexts, as when seeking authorization to commercialize some product that has been tested in an experimental trial, the submission must take a strict form and supply all evidence available. Guidelines are generally available on request from the appropriate authorities.

### *5.4.2 Scientific journals*

If the study is conducted as scientific research, the customary procedure is to attempt to publish it in an appropriate refereed scientific journal. You will have to make certain choices as to the form:

- One major publication or a series of small articles.
- A relatively general or a specialty journal.

Your choice should depend on the nature of your study and on the type of audience it may be expected to interest.

Such publication will enable your results to be rapidly disseminated throughout the world. Because such papers are refereed by peers, such publication implies that the results are scientifically sound and worthwhile. Unfortunately, reality is far from this ideal, often because of defects in design or analysis of the study not detected by this quality control process.

This form of publication is preferable to having your study appear in an edited collection of articles in book form. The latter will be less widely distributed, and often difficult for interested readers to locate. On the other hand, when the subject of a study is of lasting interest, not requiring urgent dissemination, a monograph form may be preferable.

Never publish interim results of your experimental trial if the trial is continuing. This can bias the responses still to come and, in any case, your final results may contradict the interim ones.

In a properly designed study, so-called 'negative' findings, of no relationship among key variables, can be as important as 'positive' ones. They do not indicate failure of a study. However, editors are generally very reluctant to publish the former. (It can be argued that they are responsible for fraud.) On the other hand, repeating a study often enough will inevitably lead to a significant result at random (for example, one in 20 times at a 5% significance level). That result will then be published, even although there is no basis to it, and the others ignored.

This publication bias is accentuated by the excess of small studies made in certain fields. The few that are published have a high chance of being false positives. Furthermore, it is often difficult to publish confirmatory studies that simply validate previous published findings.

One of the driving forces of scientific research is the resulting prestige for the successful research workers. At the same time, much research published in this way has no immediate concrete consequences and its validity is difficult to check. Consciously or unconsciously, these factors can influence both the decision to publish and the opinions of referees.

### 5.4.3 Oral presentations

Talks may be required in order to present your results to the sponsors, as well as at scientific meetings. Their advantages over a written report include:

- providing direct contact between you and the public;

- transmitting enthusiasm for the research;
- allowing questions, answers, and discussion.

Presenting a good talk is a matter of practice. It requires:

- confidence and enthusiasm from the beginning;
- proper planning and timing, perhaps with rehearsal;
- clarity and accuracy;
- emphasis of key points, including conclusions and recommendations;
- appropriate supporting materials.

Talks usually are limited in time so that you can only present the basic procedures and major results. Unfortunately, this leaves wide scope for bias.

For most audiences, omit the technical details of your statistical modelling, or leave them to the question period. Your presentation should cover:

- specification of your problem;
- your approach;
- your results;
- the implications.

You will have to omit or briefly summarize your detailed findings. Discussion of the implications should relate specifically to the interests of the audience.

Visual aids should highlight your key ideas, not distract the audience by their technical complexity. They are meant to help to capture and hold the audience's attention to your ideas – the message, not the medium. Arrive early and check that you can use the equipment, such as the projector and microphone. Have a back-up plan in case of equipment failure, either before or during your presentation.

Show the main points of the presentation by means of an overhead projector, a slide projector, or the projection of a computer screen. Slides, computers, and videos may look more high-tech and professional but they generally require a dark room that cuts down contact between you and the audience and may put them to sleep.

To look professional, do not use handwritten material. Between 25 and 30 transparencies would be the maximum for an hour's presentation. Each should contain minimal, but key, information, *large enough to be read.* It should allow the audience easily to follow the argument, to grasp your point, and to focus on key ideas that you can then explain in more detail. Do not read the material on the transparencies to the audience. Take care not to get in the way and block the projection of the transparencies onto the screen. Leave each transparency long enough for people to study it. If you must provide additional explanations on a board, write slowly, using large, clear characters.

Avoid tables of numbers in favour of more graphical summaries of the data. Make sure that they fit on the screen, that the axes, symbols, and legends are readable at the back of the room. Do not assume that a graph is clear to the audience; describe what you are displaying and point out obvious patterns.

If you are not an experienced speaker, practice the talk, perhaps in front of a mirror. This will allow you to:

- identify peculiar hand and body motion;
- isolate difficult language problems requiring careful phrasing;
- time the talk (although it will always take longer than you think).

Taping the presentation in front of friendly critics may also be useful to discover strange speech habits. The critics can tell you which parts are unnecessary or unclear. Listen to professional speakers to discover what effective methods they use.

Start the talk with a clear voice and a confident stance to command the attention of the audience. Be calm and friendly. *Do not apologize for being a poor speaker.* Speak slowly. Vary the level of your voice and the rate of speaking to emphasize different points appropriately and to retain the attention of your audience. Project your enthusiasm for the material and your pride in your work. Watch the audience to see how they are reacting and adjust you presentation accordingly. If they look puzzled, slow down and explain; if they are bored, skip through to something more interesting.

If you find yourself running out of time, do not speak faster. Instead, abandon large portions of the material, according to a plan prepared ahead of time. Lack of time may happen because of:

- your poor preparation;
- questions during your talk;
- bad planning on the part of the organizers;
- previous speakers running over their time.

Finish early or on time, even if you must start late. Even if the chairperson does not cut you off, the audience will react negatively.

If there is time for questions, listen attentively and take notes if necessary. Remain friendly, even to hostile questions. Repeat the question to make sure that you have understood and that everyone has heard. Answer briefly and directly. Humility is usually appreciated during questions and discussion. After the talk, seek out criticism that may aid you the next time.

You know and understand your material; the audience does not. Do not read or memorize the presentation, but use what is on the slides as prompts. Avoid jargon and technical words. Allow time for questions, either during the talk or at the end. Always remember that the audience will not be familiar with the specific work that you have done, so that you must include many apparently obvious details. Talks more often are too complex than too simple. Know your audience and adapt the presentation to them.

## 5.5 Summary

Both decision making and scientific inference are complex processes. Good statistical methods can aid in both, but are not sufficient in themselves.

Once a study is completed, you will need to evaluate it critically; the people who will read your reports certainly will. Many common statistical procedures are badly misused; take care with these. Check that the conclusions you will draw are compatible with the design that you used. Even although you did not use an experimental trial, you may be expected to draw causal conclusions. Be prepared.

You should follow the basic structure of either a technical report or a scientific paper so that your results can be clearly understood.

Many studies are only for internal use in the organization for which they were carried out. Nevertheless, if your results are important, try to disseminate them more widely. Technical reports, scientific papers, monographs, and oral presentations are the principal means available. Choose the appropriate ones for your material and audience.

Good research methodology, from initial design to final presentation of conclusions, is essential in order to protect against criticism and to make a contribution to the body of scientific knowledge that may have some potential for the advancement of human welfare.

# Bibliography


Abelson, R.P. (1995) *Statistics as Principled Argument.* Hillsdale, NJ: Lawrence Erlbaum Associates.

Altman, D.G. (1991) *Practical Statistics for Medical Research.* London: Chapman & Hall.

Altman, D.G., Gore, S.M., Gardner, M.J., and Pocock, S.J. (1983) Statistical guidelines for contributors to medical journals. *British Medical Journal* **286**, 1489–1493.

Becker, R.A. and Keller-McNulty, S. (1996) Presentation myths. *American Statistician* **50**, 112–115.

Chatfield, C. (1995) *Problem Solving. A Statistician's Guide.* London: Chapman & Hall.

Cleveland, W.S. (1985) *The Elements of Graphing Data.* Pacific Grove, CA: Wadsworth.

Cochran, W.G. (1965) The planning of observational studies of human populations. *Journal of the Royal Statistical Society* **A128**, 234–265.

Cook, T.D. and Campbell, D.T. (1979) *Quasi-experimentation. Design and Analysis Issues for Field Settings.* Boston: Houghton Mifflin.

Cox, D.R. (1958) *Planning of Experiments.* New York: John Wiley.

Doll, R. (1998) Uncovering the effects of smoking: historical perspective. *Statistical Methods in Medical Research* **7**, 87–117.

Ehrenberg, A.S.C. (1982a) *A Primer in Data Reduction.* New York: John Wiley.

Ehrenberg, A.S.C. (1982b) Writing technical papers or reports. *American Statistician* **36**, 326–329.

Elwood, J.M. (1988) *Causal Relationships in Medicine. A Practical System for Critical Appraisal.* Oxford: Oxford University Press.

Fisher, R.A. (1958) *Statistical Methods for Research Workers.* Edinburgh: Oliver and Boyd.

Fisher, R.A. (1966) *The Design of Experiments.* Edinburgh: Oliver and Boyd.

Freeman, D.H., Gonzalez, M.E., Hoaglin, D.C., and Kilss, B.A. (1983) Presenting statistical papers. *American Statistician* **37**, 106–110.

Granquist, L. (1997) The new view on editing. *International Statistical Review* **65**, 381–387.

Greenfield, T. (1996) *Research Methods. Guidance for Postgraduates.* London: Edward Arnold.

Groves, R.M. (1989) *Survey Errors and Survey Costs.* New York: John Wiley.

Hayes, R. (1998) Design of human immunodeficiency virus intervention trials in developing countries. *Journal of the Royal Statistical Society* **A161**, 251–263.

Hill, A.B. and Hill, I.D. (1991) *Bradford Hill's Principles of Medical Statistics.* London: Edward Arnold.

Huisman, M., Krol, B., and van Sonderen, E. (1998) Handling missing data by re-approaching non-respondents. *Quality and Quantity* **32**, 77–91.

Lilienfeld, D.E. and Stolley, P.D. (1994) *Foundations of Epidemiology.* Oxford: Oxford University Press.

Lindsey, J.K. (1995) *Introductory Statistics. A Modelling Approach.* Oxford: Oxford University Press.

Manly, B.F.J. (1992) *The Design and Analysis of Research Studies.* Cambridge: Cambridge University Press.

Moser, C.A. and Kalton, G. (1979) *Survey Methods in Social Investigation.* Aldershot: Dartmouth.

Mosteller, F. (1980) Classroom and platform performance. *American Statistician* **34**, 11–17.

Pocock, S.J. (1983) *Clinical Trials. A Practical Approach.* New York: John Wiley.

Senn, S. (1993) *Cross-over Trials in Clinical Research.* New York: John Wiley.

Senn, S. (1997) *Statistical Issues in Drug Development.* New York: John Wiley.

Staquet, M.J., Hays, R.D., and Fayers, P.M. (1998) *Quality of Life Assessment in Clinical Trials. Methods and Practice.* Oxford: Oxford University Press.

Tufte, E.R. (1983) *The Visual Display of Quantitative Information.* Cheshire, CT: Graphics Press.

Yates, F. (1981) *Sampling Methods for Censuses and Surveys.* London: Charles Griffin.

**More technical texts**

Barnett, V. (1991) *Sample Survey Principles and Methods.* London: Edward Arnold.

Box, G.E.P. (1966) Use and abuse of regression. *Technometrics* **8**, 625–629.

Checkoway, H., Pearcc, N.E., and Crawford-Brown, D.J. (1989) *Research Methods in Occupational Epidemiology.* Oxford: Oxford University Press.

Cochran, W.G. (1977) *Sampling Techniques.* New York: John Wiley.

Cochran, W.G. and Cox, G.M. (1957) *Experimental Designs.* New York: John Wiley.

Kelsey, J.L., Thompson, W.D., and Evans, A.S. (1986) *Methods in Observational Epidemiology.* Oxford: Oxford University Press.

Kempthorne, O. (1952) *The Design and Analysis of Experiments.* New York: John Wiley.

Kish, L. (1965) *Survey Sampling.* New York: John Wiley.

Lessler, J.T. and Kalsbeek, W.D. (1992) *Nonsampling Error in Surveys.* New York: John Wiley.

Meinert, C.L. (1986) *Clinical Trials. Design, Conduct, and Analysis.* Oxford: Oxford University Press.

Schlessselman, J.J. (1982) *Case–control Studies. Design, Conduct, Analysis.* Oxford: Oxford University Press.

# Index